Library of Congress Catalog Card Number: 89-90486

ISBN 0-9625049-0-4

Printed in the United States of America by
Press Printing Company
Fort Myers, Florida

ADDITIONAL COPIES are available by sending a check for $20, payable to Banana Book, and mail to: Banana Book, 19431 South River Road, Alva, FL 33920, USA. Add $3 extra for overseas postage.

Contributors wishing to submit new banana stories may send their articles to the above address.

Cover design by Clyde S. Stephens. The cover photograph is from an original print given to the author by John Turchick in the 1960's just before his old United Fruit Company office was razed at Pier 3, North River, New York City. The artist is unknown but the scene is Guatemalan.

To Phyllis, Linda and Carl
and the rest of the great family of *bananeros*
who have made my life in the Tropics so rewarding.

BANANEROS IN CENTRAL AMERICA

by

Clyde S. Stephens

and 27 other contributors

True Stories of the Tropics

History and Anecdotes of a Bygone Era

CONTENTS

PREFACE

Many books and articles have been written about the United Fruit Company — some pro and some con. *Conquest of the Tropics* by Frederick Adams glorified the greatness of the Company during the Golden Age of Bananas at the beginning of the century. In contrast, Kepner and Soothill did the opposite in *The Banana Empire.* Since then, the leftists have fed on the bitter fodder of banana peels, whereas those to the right have savored the sweet fruit, better known in the trade as "Green Gold." Thus, these two images are visualized as life in the "Banana Republics," depending on one's persuasions.

But there is another side that has never been told and that is what this book is about — tales of the Tropics by real banana people, *los bananeros.* There have been thousands of them, both people and their stories, beginning in the 1870's when the first commercial banana business started. With a setting in the remote and wild corners of the Tropics, story-telling of personal experiences was a way of life and many adventurous tales became legendary.

My first adventure came soon after graduating from the University of Florida when I came to Honduras to start my career with the United Fruit Company. Today, I look at the DC-3's still flying in Honduras and wonder if one of them was the same plane I boarded in Miami in early 1959. On the way to Honduras, the flight stopped in La Havana, Cuba, where Fidel Castro's brother, Raul, came aboard in full guerilla regalia, complete with beard and jungle fatigues, to begin the new revolutionary government's first diplomatic mission in Central America. Castro was accompanied by four *Cubanos* dressed in fine suits. We landed on the grass field at the international airport near the edge of San Pedro Sula. I'll never forget the huge cloud of dust as we landed and the blast of hot air as I got off the plane. There were photographers taking pictures of Castro and I felt apprehensive in the push and shove. Soon, Bob Bullock easily spotted the new *gringo* entomologist in the crowd and started me on my way. I had planned to stay only a year or so, but before I knew it, three decades had flashed by.

One of the intrigues of living in banana zones is listening to favorite stories of the old-timers. I came at a time when there were plenty of the Old Guard still around telling incredible tales of building banana ports, railroads, company towns, and new plantations in the virgin wilderness which always seemed to be in the remote corners of Central America. Much of this development was done or supervised by foreigners, mostly Americans, who had skills needed to develop and manage giant banana divisions. Despite culture shocks and inconveniences, banana pioneers found ways to survive in the harsh environment by socializing and competing in sports within what was known then as *La Zona Americana*.

Today, *La Zona Americana* and the old lifestyles of the foreign community have disappeared. As citizens of host countries became more educated and skilled, foreigners were replaced by nationals who now run the tropical divisions with few exceptions. Cable television, automobiles and private family life have replaced the gregarious environment of the old Fruit Company communities and isolated outposts. Thus, this book is dedicated to the preservation of that bygone chapter known only to a few remaining old *bananeros*.

Juan Macaya Lehmann, a Costa Rican engineer, used to tell the most incredible stories about Golfito when that banana division first started in the 1930's. It was the shock of his sudden death that made me realize that these exciting and historic accounts had to be written down before they were forever lost to the grave. In 1980, several hundred invitations were mailed to employees and retired banana people to write down their favorite stories. Thanks to their responses, this book was made possible.

The old days are gone but there is still a rich cultural heritage among Fruit Company people today. Only recently, I noted the diversity of nationalities at a small dinner party at our house honoring a visiting *bananero*. Within this group of close friends, there were guests from Guatemala, Honduras, Costa Rica, Panama, the Philippines, Canada, Scotland and the United States. All of us had vast experiences deeply rooted in the Tropics and the banana business — common denominators which have forged life-long friendships.

I wish to thank those who responded so generously with stories and photographs. Larry Mahan, a remarkable historian, was very helpful with the photographs. Special thanks are due to my wife, Phyllis, for putting the material on computer diskettes so I could edit the stories. She has also stood by me during long hours of work on this project. I am also grateful to Tam Syme, John Dickson, Carl R. Stephens and others who made valuable reviews and criticisms and to Roger Swain for his professional advice. Lastly, I appreciate the encouragement from so many *bananeros* to put this book together. The publication stands as a memorial to those who labored in the Tropics and enjoyed the banana business as much as I have.

Flying Machines and Blowdown Machines

By

Patrick C. Butler

First Class to Quepos

Occasionally, as I sit in plush comfort aboard a quiet jetliner, my mind wanders back to some of the less desirable aircraft on which I have been privileged to fly.

There have been a bunch of them, from the tri-motor Fords that Taca Airlines used to fly in and around Honduras in the 1940's to the one and only pogo stick Piper of Vargas International Airlines. The latter used to operate in any kind of weather between Puerto Armuelles and David, Panama. But one thing about Captain Vargas — he never charged you until *after* you had landed safely!

The trip that I will never forget, however, was from San Jose to Quepos, Costa Rica, sometime in 1945. The pilot was an old-timer named Sam Penry, a former crop duster from the days when the United Fruit Company was trying to control Sigatoka by aerial dusting with different fungicides. When the Company went to the fixed-pipe Bordeaux spray system, Sam and his crop duster airplane became obsolete. Afterwards, the Company introduced a freight service using Sam and his airplane between San Jose and the west coast divisions of Costa Rica at Quepos and Golfito. The plane carried fresh vegetables and spare parts from San Jose and brought back anything that anybody chose to send, from bananas to fresh fish to broken-down equipment.

The plane, a German-built Boesch, was canvas-covered, with two engines under the high wing and a fixed undercarriage. It had an open cockpit, which could be reached only from the outside, and a compartment which had formerly carried the fungicide. A single door in the side of the plane led to the compartment where the only source of light was from a small round window in the door. The normal loading process consisted of cramming all cargo in the forward part of the compartment and

---

Patrick Colin Butler was born in 1932 in Kingston, Jamaica, where his British parents were on assignment for the United Fruit Company. He attended schools in the Tropics, in England, and at Cornell and Illinois. After getting his doctorate from North Carolina, Pat married Jacqueline Rogers and moved to Honduras in 1957 as a soils chemist. Later, as an executive, he headed research for the United Fruit Company in Boston and New York. Pat died in Miami in 1985.

Sam Penry's converted crop duster flew banana people and their cargo between San Jose, Costa Rica, and the Golfito and Quepos plantations in the 1940's. (Courtesy of Richard Washburn and Fritz Stargardter)

securing it with a metal grid. If there was any space left, a small wooden bench was placed along each side of the remaining space and this became the passenger cabin. If you really wanted to go to Quepos, this was where you rode, unless Sam asked you to join him in the open cockpit. However, flying as copilot was not always pleasant because Sam Penry chewed tobacco and was famous for spitting tobacco juice during flight.

In 1945 I was around 13 years old and was very much in love with the girl next door in San Jose. Her father and my father, Alfred P. Butler, both worked for the Fruit Company and there was a fair amount of socializing between the two families. So I was quite pleased when it was decided that we should all go to Quepos for a couple of weeks vacation between school terms. I looked forward to the trip with great anticipation and showed up at the airstrip on the day of departure wearing my very best trousers and sporting a brand new haircut. Of course, all this was pure show because I was too shy to do any more than say hello to the girl!

Anyway, we did stuff ourselves into the plane. I ended up on the very end of the bench next to the door and was waiting for them to close the door and take off, when a calf's head appeared in the doorway. Somebody outside the plane said, "Would somebody in there grab this rope and tie it to the barrier? Then we'll push her in!"

Well, by Jove, they pushed that calf all the way in and secured her head to the barrier up front so that she was standing in the aisle with her rear end right over my knees. By this time, we were all packed in so tightly that nobody could move. As we took off, I had two horrible thoughts: one was that that old relic of an airplane would never make it off the runway, and the other was that the calf might just be as scared as I was. I was wrong about the plane but dead right about the calf! Since I couldn't move my legs I had to watch fascinated as the tail went up and she delivered right onto my brand new gabardines.

It didn't help that the plane was slow and, of course, totally lacking in ventilation. My embarassment was extreme by the time we landed and it took several days for me to be able to face anybody. The pants didn't survive the treatment and had to be discarded. But the worst part of it was that everybody else on the plane thought it was all very funny, except for Sam Penry, who was mad about his plane getting messed up.

Needless to say, I have made it a rule on all flights since then to make certain there are no livestock riding in the passenger compartment near me.

The Blowdown Machine

As is well-known by all banana men, there are usually more banana losses from wind storms, known as blowdowns, than from all the banana pests and diseases put together. This was even more pronounced in the days of the tall Gros Michel variety.

In 1958, Dr. Jesse Hobson, then Vice-President of Research for the United Fruit Company, decided that the Research Department in La Lima, Honduras should look into this problem to determine if there was any practical way of reducing blowdown losses. Since this was to be a research project, the first thing we had to do was hold a staff meeting to define the problem and evaluate different approaches. Windbreaks had been tried before and were ineffective. Instead, various ways of strengthening the plant itself were considered. These included everything from multiple propping of individual plants to metal collars around the pseudostem trunk. However, nobody could describe the exact sequence of events which occurred as the wind hit the plant. This would have required that an individual be present at the precise instant that the wind storm developed and was able to observe the plant's bending and twisting, preferably in slow motion.

Blowdowns, however, occur sporadically and without advance warning. Usually we observe the results, not the actual happening. It would be quite impossible to run any kind of controlled experiment if we had to depend on chance occurrence of blowdowns.

The logical conclusion was, therefore, to have a system whereby we could create blowdowns *on demand.* As a result, we built what I believe was the only blowdown-causing machine ever seen in the world-wide banana industry. This machine, when finished, consisted of four powerful aircraft engines mounted on a vertical frame above an axle with two rubber-tired wheels and a hitch so it could be pulled by a tractor to the location of the programmed blowdown. A control panel mounted on the frame enabled the operator to run any combination of the engines at different speeds. The idea was that the machine would be placed in front of a small stand of bananas and a movie camera would then record everything that happened. Later, the movie could be studied for details.

It should be noted here that relations between research and production people are, at the best of times, a little strained and, at other times, downright antagonistic. In the 1950's, we still had some of the original banana men who had worked for Sam Zemurray in the days of the Cuyamel Fruit Company and whose attitude toward research was, at best, barely tolerant. Their reaction to the creation of a blowdown machine was predictible. They had always known that research people were weird; the fact that we now wanted to go out and deliberately blow down bananas with a machine confirmed their every suspicion. With each highball consumed at the Golf Club, the issue of the blowdown machine became more heated and the sarcasm accumulated.

Under these conditions, it was considered prudent not to try out the machine but store it away until everybody had forgotten about it. As a "research tool," it was actually never used. Dr. Hobson left the company and his successor, Bill van Diepen, carefully avoided bringing up the subject again.

The machine actually did get to blow down some bananas before it was finally dismantled and forgotten. The Public Relations Department from the Boston office was producing a film about the company entitled *Treasure of the Tropics* and the script called for a brief scene showing a blowdown. Somebody remembered the machine and it was hauled out to the field for what proved to be its one and only performance. It did the job admirably — all the bananas blew down and the action was recorded on film.

The last time I saw "Old Blowndown," she was parked in the tall grass at the Campo Dos airfield outside La Lima. The aircraft engines had long since been removed and most people had no idea that the skeleton in the bushes was all that remained of a truly unique machine.

The Green Gold of Bocas del Toro

By

Lawrence Mahan

During the latter half of the 1880's, Michael Theodore Snyder, better known as M.T., decided that adventure and the promise of success would not be found in New York City as he had originally thought. Instead, the lure of excitement was greater in California. With his recent completion of engineering courses at college, these services would certainly be in demand in gold country. Contemplating further, M.T. thought that even if his abilities were not needed, finding gold would surely remedy any financial situation.

While living in New York with his brother, Charles Louis, also a graduate in engineering, a letter arrived from John Palmer, an old family friend. Mr. Palmer lived in Bocas del Toro, Panama, then in the Republic of Colombia, where he operated a large farm on the island of Bastimentos. He encouraged the Snyder brothers to come to Panama and start banana plantations. The market looked promising and John knew that someone with the Snyders' engineering abilities would be needed to operate and develop any large plantation systems.

M.T. and Charles talked it over. After conferring with the Tilton grocery firm of New York City regarding financing of this endeavor, M.T. agreed to go to Panama and see if the project looked worthwhile. There is evidence of a connection between the Tiltons and Carl B. Franc, pioneer banana shipper from the Colon terminal of the Panama Railroad, which may have influenced young Snyder into going to the Tropics. A steamship ticket was purchased and M.T. was soon on his way.

Once Snyder arrived in Colon, the smells and sounds of this new environment renewed his excitement for adventure. Sev-

---

Lawrence Mahan was born in 1937 and raised in Melrose near Boston. Because of his driving academic curiosity, a lost chapter of banana history was saved from extinction. The search for the Snyder banana story began when he started building a ferro-cement boat, sixty feet long, in his back yard in 1974 in Marstons Mills on Cape Cod. This craft, modeled after a 1767 British schooner, has a seven-ton Wolverine diesel engine. The Snyders bought the Wolverine Motor Works with money from the sale of their banana company in Panama at the turn of the century. Larry and his wife, Marlene, plan to drop anchor in Bocas del Toro, Panama, where the Snyder banana story began in 1890. Meanwhile, Larry is a master mechanic at his shop in Hyannis.

eral weeks passed before M.T. could arrange passage on a small schooner to Bocas del Toro. During the interim, he examined the ongoing French canal effort which was already on the decline primarily due to bad planning and corruption.

The visit to John Palmer's farm on Bastimentos Island convinced M.T. to make a quick survey of available lands suitable for banana farms. As it turned out, much of the land around the Chiriqui Lagoon was suitable for banana growing. There was also the benefit of quiet, deep water for steamships within the lagoon where loading operations could be handled safely and quickly for transport to New Orleans, Mobile and other ports. M.T. sent a long letter to his brother Charles outlining the virtues of the banana idea and explaining how easy it would be to acquire the needed labor force for fruit production. Charles soon negotiated the required capital and joined M.T. in Bocas del Toro.

A large tract of inexpensive land was acquired along the Cricamola River as well as some land in the Chiriqui Grande area. As soon as both areas were surveyed, the first banana plantations were started.

Soon, arrangements were made for more land bordering the Chiriqui Lagoon at Cabbage Creek (Fish Creek), Robalo River, Pumpkin River, Caucho River and the mainland in front of Monkey Key. Headquarters and homes were established at the Cricamola River plantation and on Monkey Key.

Help was recruited from the local population. Also, migrant Jamacian blacks came to work in the banana plantations rather than keep on with the floundering French Canal. To house the workers, a large network of huts with kitchens and other buildings were constructed at the larger plantations. Sheds for banana lighters and launches were built. With all this work and money, it became apparent to the Bocas del Toro region that the Snyders were a boon to the economy.

M.T. fell in love with Seveline Matthews who lived on the farm next to John Palmer's property. Seveline was the daughter of Mary Jane Shepard and William Matthews, owners of Utica Farm on Bastimentos Island. These were Caribbean people of Anglo decent. On May 20, 1890, M.T. and Seveline were married at Utica Farm and moved into a home on Monkey Key.

Charles realized that the human labor required to row or pole banana lighters up and down the various rivers was tedious and oftentimes dangerous. He had seen examples of the infant

Charles Louis Snyder (left) and Michael Theodore Snyder (third from left) and their brother, Joseph Alfred, were the first major planters in Bocas del Toro, Panama, then still a part of Colombia. During the 1890's they developed plantations along rivers flowing into the Chiriqui Lagoon. (Courtesy of Louis Irving Snyder, Jr. and Larry Mahan)

Snyder Brothers banana loading operation on the Cricamola River, Panama, then part of Colombia, about 1894. A small locomotive powered by a Wolverine gas engine pulled bananas over a narrow gauge tramline to the loading conveyor. (Courtesy of Louis Irving Snyder, Jr. and Larry Mahan)

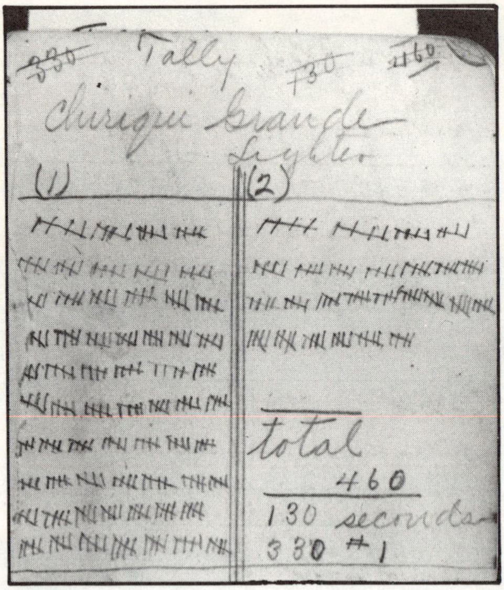

Tally of first and second class banana bunches at Chiriqui Grande, Panama, then Colombia, in 1894. From the notebook of Charles Louis Synder's son, Louis Irving Snyder, who worked with Snyder Brothers during his school vacations. (Courtesy of Louis Irving Snyder, Jr. and Larry Mahan)

internal combustion engine industry when living in New York and began to search for a good engine company that was building a product that could survive in this remote site. As a result, the famous Wolverine Motor Works of Grand Rapids, Michigan, was contacted. Before long, two launches were delivered to Bocas del Toro, each equipped with a two-cycle, one-cylinder, 9 horsepower engine.

The task of hauling bananas and personnel in this water-oriented region quickly eased with the introduction of these powered craft. This resulted in more orders for additional launches and Wolverine engines. By early 1895, eight Wolverine-powered craft were pushing small barges loaded with bananas to the town of Bocas del Toro for subsequent loading onto steamships.

At this time, Charles decided to expand the Cricamola River plantation further into the jungle. The increase in area required a larger group of laborers and horses to transport bunches of bananas to the loading sites along the river. Realizing how well the motor-powered launches increased production with less manpower, Charles envisioned a small railway system throughout each plantation. Bananas could be loaded onto flat cars and each car would carry many times that of a man and horse.

During 1895, Charles arranged a trip to the Wolverine Motor Works at Grand Rapids. The visit resulted in Charles colaborating with the company's own engineers on ideas for the locomotive. A gasoline-powered, narrow-gauge locomotive was designed utilizing a two-cycle, single-cylinder engine of the type already powering the banana launches. This kind of standardization would simplify repairs and maintenance.

Charles stayed in Grand Rapids for several months to oversee the construction of the locomotive and cars and learn machine shop techniques which the Snyders would use in Panama. The future looked brighter now that modern technology was coming to Bocas del Toro. Once Charles arrived back in Bocas del Toro with the locomotive, cars, and other machine shop equipment, the Snyder brothers installed the railroad at the Cricamola River plantation. A small machine shop for repairs was set up at the plantation and banana acreage was soon expanded along the new tracks.

Another larger machine shop was built on Monkey Key and more lands were purchased on the adjoining mainland. These included lands formerly owned by Mr. Matlock and another piece owned by Mr. Reuterios. More railroads were planned for the plantations in front of Monkey Key and at Pumpkin River, Robalo, and Chiriqui Grande. The Monkey Key machine shop would serve to repair the launches as well as future railroad equipment.

The business of the Snyder brothers continued to grow at a very accelerated rate. Charles and M.T. decided to convince another brother, Joseph Alfred, better known as J.A., then living in Nebraska, to move with his family to Bocas del Toro and become part of their growing concern.

Offices were set up in the towns of Bocas del Toro, Colon, and Mobile, Alabama. Charles built a warehouse in Bocas del Toro along the sea for supplies and a subsequent import/export business. A Wolverine agency was also established at the warehouse due to interest and use by other companies and individuals in the area. Bananas were shipped out on chartered steamships. The returning steamships would bring hard goods from the United States for sale at the Snyder warehouse. Stores in town would buy some of their stocking goods from the Snyders. These voyages to and from the United States became profitable both ways and the business generated attracted many new people to Bocas del Toro. This caused the economy of the region to prosper.

Charles' family, consisting of his wife Zerauh and a son

Louis Irving, lived in New York as Louis was still in school studying law. When school was out, they both would come to Bocas del Toro. Louis helped with administrative duties which included some field work doing tally counts of banana bunches.

Personal profits allowed each brother to invest in other schemes. Charles bought shares in the growing Wolverine Motor Works. M.T. purchased land in the Changuinola area for cattle raising and future banana plantations. J.A. banked his money and interested his in-laws in the idea of starting a fruit import/export business in New Orleans. The yellow gold M.T. so excitedly sought to find in California had turned into bananas — the green gold of Bocas del Toro.

During the latter part of 1898, Minor C. Keith, the famous railroad builder and banana tycoon in Costa Rica, bought into the Snyder brothers' company. Keith's investment was strong enough to allow him to become a deciding factor in incorporating the company. The name was changed from Snyder Brothers to the Snyder Banana Company and was duly incorporated in New Jersey on December 21, 1898. By early spring of 1899, the Snyder brothers agreed to sell their shares in the Company to the newly formed United Fruit Company. This deal was finalized on May 21, 1899.

For a time after the sale, all the Snyder brothers stayed on in Bocas del Toro in order to help with the transition of ownership to the United Fruit Company. Once all the field problems were worked out and most of the original personnel were assimulated into the new company, the Snyders went separate ways.

Charles, the eldest, purchased a controlling interest in the Wolverine Motor Works and moved with his family directly to Grand Rapids to operate the Company. M.T. and J.A. also purchased some shares in Wolverine but did not become physically involved with the business.

J.A. continued to work with M.T. but based his own operations in New Orelans where he had established a fruit import/export business with his in-laws, the Niebergall family. Several years later, J.A. joined M.T. in projects in Guatemala and Mexico while still operating at New Orleans.

M.T. stayed to operate the store and supply warehouse in the town of Bocas del Toro. A fine new home was also built for his own growing family. His plan was to enlarge his cattle development in Changuinola and also sell bananas to the United Fruit Company.

Changuinola was virgin territory and had no port facility. Bananas from this new area could not be conveniently transported by barge through the dangerous mouth of the Changuinola River and down the coast through the rough open sea. Consequently, Michael T. Snyder started constructing an eight-mile canal that eventually would connect the Changuinola River with the deep water at Boca del Drago in the Almirante Bay.

Construction of the canal was started at a quiet cove just south of Donato Point on Almirante Bay. Although a large swamp encompassed the area all the way to the Changuinola River, M.T. thought the dredge work would be relatively easy. However, after months of work, it became obvious that the unstable mire of the swamp was filling back into the already dredged section of the canal and this route had to be abandoned.

It was decided to relocate to firmer ground just behind the Caribbean beach. This route paralleled the sea shore all the way to the Changuinola River, cutting a line between the sandy beach and the swamp behind. Since the entrance to the new waterway was closer to the open pass at Boca del Drago, wave action caused some sedimentation at the mouth of the canal which required occasional dredging.

The United Fruit Company quickly expanded all the former Snyder plantations around the Chiriqui Lagoon. Needing more land to plant, they offered M.T. large financial inducements to sell his remaining banana lands in Changuinola. The Company offered to purchase all of M.T.'s land, buildings, livestock, and goods for the sum of $500,000 in gold if he completed the Changuinola Canal by 1903. The canal and associated lands, along with all dredging equipment, were to be included in this agreement. M.T. continued to work on the canal and after its completion in 1903, the sale to the United Fruit Company was finalized with the documents being signed by Minor Keith, president of the Company. That same year, Panama became independent from Colombia.

Adventure and new frontiers were still paramount on M.T.'s mind. During the next fifteen years, he tried various schemes which included gold mining in Nicaragua, railroad work in Guatemala, and involvement with bananas in Guatemala and Mexico. During part of this time, he left his large family in Bocas del Toro. In time, the children were sent to Europe and educated at an early age in boarding schools in Scotland, England and France. Several of the Snyder children married and remained in Europe. Only one daughter, Ella Olla, returned with her mother, Seveline, from Europe to Bocas del Toro and married a Panamanian, a Sr. Bendiburg.

Ultimately, all three Snyder brothers finished their life's work in the United States. The Wolverine Motor Works was the only Snyder-owned company that survived. During 1955, this company was purchased by the Coulter McKenzie Machine Company and by 1957 no further engines were built, although spare parts were supplied and repairs performed as late as 1972.

In 1984, this author purchased the remains, excluding real estate, of the Wolverine Motor Works for the expressed purpose of saving this unique piece of history.

Banana Railroads

By

Richard Yudin and Clyde Stephens

Banana railroading began in the 1870's when young Minor C. Keith, the builder of the Costa Rica Railway from Limon to San Jose, began planting Gros Michel bananas along an uncompleted section of tracks near the Caribbean. To expand to new plantations, Keith built a controversial private railway to Limon which was called the Northern Railroad. Elsewhere, by the late 1890's, he controlled plantations and railroads in Panama at Bocas del Toro (then a part of Colombia) and at Santa Marta, Colombia, where bananas were exported to Mobile and New Orleans. The bankruptcy of his broker in New Orleans forced Keith to join a competitor, the Boston Fruit Company, which resulted in the formation of the United Fruit Company in 1899. With this merger, the plantations and railroads formerly owned by the Boston Fruit Company in Cuba, Jamaica, and the Dominican Republic became part of Keith's empire in Central and South America.

Keith's Northern Railroad, which then became part of the United Fruit Company, leased the Costa Rica Railway in 1904 to expand banana plantings. Down the coast across the border in Panama, the pioneer Snyder Brothers had been bought out and a railroad was started in 1906 from the new port of Almirante. By 1909 this new line extended across huge bridges to Changuinola, both sides of the Sixaola River and, a few years later, far up the Talamanca Valley in Costa Rica, some 80 kilometers distant from the port.

The Santa Marta Railway in Colombia was begun by British interests to serve sugar cane plantings in the 1880's. However, a decade later, Keith took over the line and planted bananas at Sevilla and Rio Frio.

Richard Yudin was born in 1950 of British parents in Montevideo, Uruguay. He studied tropical agriculture at Cornell University and became interested in the United Fruit Company. Richard joined UFC in Golfito, Costa Rica, in 1973. Because of his keen interest in railroading as a hobby, he wrote an outline of United Fruit Company railroad history. Later, he was transferred to Davao, Philippines. Since Richard left UFC, he received an MBA in Manila. His wife is from the Philippines. After working for Standard Fruit Company in Costa Rica and Colombia, the family moved to London where Richard is financial director of an essential oils company.

The Changuinola Railway in Bocas del Toro, Panama, was developed during the first decade of the 20th century. Before the Changuinola River bridge was completed in 1908, locomotives and rolling stock arrived in pieces via the Snyder Canal and were assembled in the Changuinola machine shop. George Perkins (third from left) worked on bridges and trains during this decade. (Courtesy of George W. Perkins II and Larry Mahan)

George Perkins (black suit) by his house in the Changuinola-Sixaola plantations of Panama in the early 1900's. These were Yankee bridge builders and railroad men from New England, as denoted by the name on the house — Nashua (New Hampshire). (Courtesy of George W. Perkins II and Larry Mahan)

After completion of the Changuinola River bridge in 1908, the Snyder Canal became obsolete and all rail traffic went directly to and from the new port of Almirante, Panama. (Courtesy of George W. Perkins II and Larry Mahan)

In the Dominican Republic, some bananas had been exported from the north coast over a short railroad east of Puerta Plata. To expand the operation and begin a new division, a complete company port town with wharf was built on the north coast at Sosua. However, because of cool weather affecting production during winter months and the risks of hurricanes, Company strategy changed and the new town was completely abandoned in 1906 before any bananas were ever shipped. This strategy also applied to the Cuban banana operation at Banes on Nipe Bay on the northeast coast and the area was soon converted to sugar cane. Rail lines were extended from Banes to another Company cane area at Preston.

United Fruit had only a few lines in Jamaica at Port Antonio, where Boston Fruit Company operated before, and at Saint Catherines. Bananas from private growers were sent on government railroads.

In eastern Guatemala, plantations began in the Motagua Valley in 1905. This fruit was hauled over farm lines of the United Fruit Company which connected with the existing Guatemala Railway Company and on to Puerto Barrios. In 1904,

Minor Keith and William Van Horn contracted to complete the main line from Puerto Barrios to Guatemala City and eventually acquired control of the network throughout the country and neighboring El Salvador. In 1912, this large rail system was known as IRCA, or International Railways of Central America. Keith's ambition was to have a rail network from North America to Panama. United Fruit's ownership and role in IRCA was always denied, although Keith controlled the railroad and was the first vice-president of United Fruit Company at the same time.

In Honduras, concessions were negotiated with the government in 1912 to build two railways. At historic Trujillo, founded in 1525, the Company took over the Honduras Lumber Company's standard gauge track and regauged it to 42 inches. Later, a Fruit Company town and port were developed across the harbor at Puerto Castilla. A long line was constructed up the Aguan Valley to develop banana plantations. This operation was called the Truxillo Railroad Company which included a railway eastward to Black River where bananas and mahogany logs were hauled to Puerto Castilla. Over the years, Panama disease caused great declines in banana production. Then, after a hurricane in 1938, the huge division was closed and much of the Puerto Castilla staff, port, buildings, railroads, bridges, steam engines and banana cars were sent to two new divisions on the Pacific coast of Costa Rica at Golfito and Quepos.

The other concession in Honduras was called Tela Railroad Company. One of the first developments was the Colorado District in the Lean River valley east of the port of Tela. The Company also penetrated west into the Sula Valley, and, by the 1920's, had planted the east bank of the Rio Ulua. Just across the river was a formidable rival, Sam Zemurray's Cuyamel Fruit Company, which extended all the way to the Guatemalan border.

The love-hate relationship between Sam Zemurray and United Fruit is well documented. Zemurray, with capital partly obtained from United, bought out a defunct banana operation at Cuyamel, Honduras, in 1906. Later, United renounced its share of the Cuyamel, assuming it would go broke. But Zemurray extended his railroad east to a small port at Omoa and prospered. He also pushed west into territory claimed by Guatemala and United Fruit in the Motagua Valley. In the end, Guatemala and United Fruit won this disputed territory. Undaunted, Zemurray leased the National Railroad of Honduras which began at Puerto Cortes and went south through San Pedro Sula to Potrerillos. Banana plantations sprang up along the main line and new branches. In 1919, a bridge was built across the Rio Chamelecon at Baracoa and permitted expansion to the west bank of the Ulua River where the competition, United Fruit, was thriving on the opposite side. Besides bananas, Zemurray controlled huge sugar cane operations at La Lima under the Cortes Development Company in the 1920's.

The Cuyamel Fruit Company also had a small banana operation in Nicaragua on the Caribbean where tramways were used to get fruit to river barges. Fruit was tugged to waiting ships anchored at Bluefields.

In 1930, Zemurray retired and sold Cuyamel to United Fruit Company in exchange for 300,000 shares of UFC stock worth $31,500,000. However, when the value of his stock plunged during the depression, he came back in 1932 in a surprise power ploy and took over the entire United Fruit Company with a vengeance. Zemurray kicked out many of the old Boston Brahmins and some United Fruit officials in the Tropics and replaced them with trusted and tough banana men from his former Cuyamel Fruit Company. This maneuver got the United Fruit Company through the depression.

With the merger of Cuyamel and Tela Railroad Company, United had virtually complete control of all the railroads and banana plantations of Honduras except for Standard Fruit Company holdings at La Ceiba. Heretofore, the Tela Railroad had never crossed over the Ulua River, but in 1933, a large bridge was built at Guanacastal, thus linking and uniting the two former rival territories.

With huge developments in so many locations, the 1920's could be considered the "Golden Age" of railroading for United Fruit. The Company had six banana divisions in full production: Bananera, Guatemala; Tela Railroad and Truxillo Railroad in Honduras; Limon, Costa Rica; Bocas del Toro, Panama; and Santa Marta, Colombia. Also, Jamaica was still producing bananas. In Cuba, there were two major sugarcane divisions at Preston and Banes. All of these divisions were on the Caribbean side of Central and South America, except for the Cuban cane operation which was on the Atlantic. Within a few years, the Company would own over 200 locomotives to pull 7,000 rail cars over nearly 1,800 miles of track to keep business moving.

In contrast, the 1930's was a rough decade. The global economic depression took its toll and times were hard. But another major factor determined the future of bananas and railroading and that was Panama disease. Because of this soil-born fungus disease, banana production dropped to a minimum in Guatemala and Honduras, and the Limon and Bocas del Toro divisions practically stopped shipping fruit altogether. As the disease wiped out thousands of acres each year, the land was planted to cacao or put into cattle to hold the property. Vast areas simply grew back to jungle again. Only Santa Marta, Colombia, was not attacked by the disease and that division continued to produce bananas.

The salvation of the banana industry was to find new virgin lands to plant. This was done by surveying huge tracts of rain forest on the Pacific side which lead to the initiation of Puerto Armuelles in western Panama in 1927. To begin with, the Company acquired 25,000 acres of mostly untouched land from a defunct American outfit called Panama Sugar Company that had a sugar mill at Progreso. The operation had gone broke during World War I and their assets grew back into jungle, including a small railroad that ended near Puerto Armuelles.

In 1928, the temporary headquarters of United's new Chiriqui Land Company occupied what remained of the old sugar company in Progreso and the first bananas were planted in that region. To launch this new project, personnel, materials and supplies, including two steam engines and rolling stock, were sent from the dying Bocas Division across the mountain and from the shrinking Limon operation in Costa Rica. Under obligation, the Company built a main rail line for the Panama government from the developing port of Puerto Armuelles to Progreso. This main line connected with the Ferrocarril Nacional de Chiriqui, a railway built in 1914 from David to Progreso for the government.

As banana farms were planted, the Company built a network of branch rail lines covering a huge block between the Rio Chiriqui Viejo and the present-day Costa Rican border. At that time, the border was still disputed, but in 1941, a negotiated boundary traversed the plantations which left several farms "stranded" in Costa Rican territory. The lines on the Costa Rican side were rerouted to a single border checkpoint at Puerto Gonzalez where fruit trains continued on to the Panama port of Puerto Armuelles.

A decade after Panama's Armuelles Division was started, United Fruit opened up two new virgin areas on Costa Rica's Pacific coast at Golfito and Quepos in 1937. The first bananas were bought from small growers in Golfito's Golfo Dulce and the Palmar region, then barged up the coast to Puntarenas, transferred to rail cars and freighted 180 miles across the country to Port Limon on the Caribbean. During this time, tram lines were built on both sides of Rio Terraba at Palmar to move fruit to barges tied up at the tidewater river port of Pozo (now Cortes). From this river port, a main line was started across the forested coastal plain. Simultaneously, the terminal end of the main line was started at the port of Golfito which was under construction. These tracks finally met in 1941 at Piedras Blancas and the first bunches of bananas from Palmar were exported from the new port after traveling over 50 miles.

Later, the Golfito line branched across the Coto Valley and continued to the Panama border checkpoint at Puerto Gonzalez, thus linking the Compañia Bananera de Costa Rica with Panama's Chiriqui Land Company. Since Panama's 36-inch gauge tracks had previously extended into Costa Rican territory before the boundary was settled, traffic from the new Golfito side came on 42-inch lines and was restricted to a limited area along the Costa Rican border. To correct this situation, a third rail was added so that Costa Rican rail traffic could run more freely on its own side of the border. In 1964, these Costa Rican border farms were transferred from the management of Panama's Chiriqui Land Company to the Compañia Bananera de Costa Rica and the border fruit was then exported from Golfito. As a result, the original "Panama" tracks on the Costa Rican side had to be regauged to 42 inches. The Costa Rican railway, known as the Ferrocarril del Sur, operated until a fatal labor strike closed down the Golfito banana industry in 1984.

At Quepos, known as "the Acapulco of Costa Rica," the first fruit was purchased from the Lutz family and other private growers, then barged to Puntarenas and shipped by rail across the country to Limon. Meanwhile, the Company built a dock, railhead and lines to new banana farms. A giant hill was blasted to get foundation material for building over swampy shoreline areas. Because of the rapid decline in production from Panama disease, the last bananas were exported from Quepos in 1956 and the division was planted into African oil palms, along with some cacao and cattle pastures. Under pressure from the government, the Company tracks were removed a decade later and roads were built over the old right-of-way.

In the late 1920's, the Company also bought land on the west side of Guatemala at Tiquisate. Since some bananas were exported from an undeveloped Pacific port for the California market during that decade, a new port was planned. However, economic factors changed the future. During the depression, export coffee and other rail freight declined so drastically that IRCA offered United Fruit a special rate to transport fruit on their railroad to the east coast in order to stay in business. Henceforth, all bananas from Tiquisate traveled some 36 hours across the country to the Caribbean at Puerto Barrios and plans for a new Pacific port were cancelled. Banana production at Tiquisate was minimal until a surge in activity in the late 1930's. Afterwards, United Fruit expanded its rail network in the new plantations.

From the late 1920's to the early 1950's, Gros Michel bananas, grown as a shade for coffee, were purchased by the Company in Guatemala. UFC banana men from Tiquisate maintained a Company guest house in the cold Guatemalan highlands to give technical assistance to coffee planters where shade bananas were bought. This fruit was also exported from Puerto Barrios. In 1964, all fruit shipments from the Pacific side stopped when Tiquisate was abandoned because of Panama disease, blowdowns from wind storms, and the long haul to Puerto Barrios.

As a result of World War II, banana production dropped sharply and railroading declined. However, the Company more than survived when its banana ships of the Great White Fleet were requisitioned into service by the British and American governments to transport refrigerated goods and war materials. Another lucrative contract was made by the American government with the UFC to grow Manila hemp, or abaca, a banana-type plant, on banana lands to produce fibers for rope needed in the war. Hemp plantations and fiber processing plants were developed in Changuinola, Panama; Limon, Costa Rica; Guaymas, Honduras; and Bananera, Guatemala. All of these operations used previously-existing railroads except for Costa Rica. There, the Company built over 250 miles of 24-inch gauge tracks in the Limon area to serve the abaca plantations.

After the war, Gros Michel bananas were replanted in Central America, especially in the 1950's, and railroads were rehabilitated. In the Dominican Republic in 1946, the Grenada Company started a new railroad in the arid Yaque del Norte Valley. The Company planted bananas at Walterio and La Cruz but most of the fruit was purchased from Dominican growers. Bunches were loaded onto a railway which extended to Guayubin and terminated at the new Company port of Puerto Libertador (Manzanillo) by the border of Haiti on the north coast. The Company left in 1963 because of poor production and political and labor unrest after strongman Trujillo was assassinated in 1961.

Production was also increased in Ecuador and the old system of barging fruit out to ships was revived. United's Tenguel District was served by a 30-inch gauge tramway which ran through the farms to a barge dock at Conchero. United also bought bananas from local producers. This division was shut down in 1960 because of labor and political problems.

These, then, were the 16 banana railroads operated at various times by the United Fruit Company. Not all of these rail systems existed at the same time because, as one area declined from Panama disease, floods, hurricanes, or strikes, assets were dismantled and reassembled in a new division. This was difficult in some cases because of different gauges of tracks, engines and rolling stock. Banes, Cuba, originally had 36-inch lines but these were widened in 1934 to a standard 56.5-inch gauge which connected with another railroad to La Havana. United Fruit kept a luxurious car in Preston for executives to travel on this line to the Cuban capital. Guatemala and Panama also used 36-inch gauges but Honduras, Costa Rica, Colombia and the Dominican Republic had 42-inch tracks. Especially in the early days, there were smaller lines with 24 or 30-inch gauges, such as the railways in British Honduras and Surinam.

The Company's first steam locomotives were the Porter Mogul. There were also various Baldwin, Alco, and Rogers models. In Santa Marta, Colombia, the entire fleet was British-built until some Baldwin 282's arrived in the 1930's.

The second generation of steam locomotives came in the 1930's and 40's and were all Mikados of three types: an 80-ton Baldwin design, a similar German Krupp product, and a 90-ton Alco originally built for the U.S. Army and bought as surplus stock after the war in 1946. The Baldwins went to Santa Marta, Golfito and Tiquisate, the Krupps to Tela, Santa Marta and Tiquisate (some of these Krupps reportedly spent World War II in

Essen, Germany), and the Alco "MacArthurs" went to Tela and Golfito. However, these engines were transferred from one division to another as they were needed. By the 1950's, most Company railroads had switched to diesel-electric locomotives. Tela Railroad was the last division with steam engines but these were phased out in the late 1960's.

The first internal combustion locomotives had Davenport gas-electric engines and were used in Puerto Armuelles. These were converted to diesel-electric. In 1929, the Chiriqui Land Company pioneered the use of six original prototype General Electric diesel-electric locomotives, along with some Alco 50-ton machines which came in 1936. The General Electrics are still in use after 60 years of service in Armuelles.

By the 1980's, banana railroading had changed drastically. In 1959, Fidel Castro seized the entire Cuban operation. The railways of Panama and Honduras were expropriated by the government in the 1970's but the banana companies still manage the operations. Bananera in Guatemala was sold in 1972 to DelMonte which still runs the banana section of the railroad. The Golfito, Costa Rica, line is a dead railroad, whereas the Limon side was electrified by the government in recent years. Colombia runs the Santa Marta line. With the coming of roads, trucking fruit to port is *de moda* in Ecuador, Colombia, Costa Rica, Jamaica and other areas, especially where the Company buys bananas.

Railroads were the backbone of the old banana and sugar cane divisions and the greatness of the United Fruit Company would never have been achieved without them. Banana railroading was a remarkable era, but like always, the scene is still changing back and forth to the future.

Luck and Itching Feet

By

Harry Hinson Hutchings, Sr.

Farewell to Jamaica

I arrived in Kingston, Jamaica in early 1920 and immediately set about getting to the Cayman Islands where my father had been transferred as Governor in the Colonial Civil Service. The family — my mother, six sisters, and a brother — I had not seen since I left the Turks Islands in 1915.

Grand Cayman, the largest of the group, lies about two hundred miles northwest of the west end of Jamaica. The fact that transportation was by sailing schooner and took anywhere from three to five days only increased my expectations of a pleasant voyage. Late one afternoon, we left Kingston with the last of the sea breezes, expecting to pick up the land breeze in the early evening to move us westward along the coast. But all that night and the next day and night, the land and sea breezes unaccountably failed, and we merely "ghosted" along the coast in full view of Jamaica's mountains. It was full daylight of the second day before we cleared Negril in a fine breeze which pushed the *Fulmar* along on her course for Grand Cayman. A grand sailing breeze, it did not fail that day and night nor the following day. I renewed my appreciation of the sights and sounds of a sailing ship at sea in a strong wind. The powerful updraft and spray as the bow plunged to the sea, the lift of the schooner to the regular swell and the downward rush as it passed — wonderful sailing, I thought, as I watched the sun come up over the windswept sea.

At noon on the third day, we hoped to be within sight of Grand Cayman, whose highest point is only fifty feet above sea level. If we missed the island and carried on to the west, the next landfall would be the coast of Yucatan in Mexico. At noon, there was nothing in sight from the masthead.

The captain of the *Fulmar,* the mate, and another schooner captain on passage, then had a conference in which it was de-

---

Harry Hinson Hutchings was born in 1896 in the Turks and Caicos Islands, British West Indies, and grew up in Bermuda and England. From 1922 to 1955, he worked for United Fruit in Honduras, Costa Rica and Guatemala where he was Assistant Manager in the Bananera Division. Mr. Hutchings retired in Sea Level, North Carolina, where he wrote his memoirs.

cided to keep on the same course westward until two p.m. when, if the island was not in sight, they would heave the ship and wait for the booby birds — yes, the booby birds — which was not as foolish as it sounds. A slight compass error, or an indifferent helmsman in the four-hour watch at night, could easily allow a schooner to get off course sufficiently to miss a fifty-foot elevation at the end of a two hundred mile run, but the boobies, or terns, returning to land after the day's fishing, are infallible. Soon after three, we saw the first birds in sustained flight in one direction. They were going home and we followed. As it happened, we had over-reached the island, but well before dark, it was in sight, thanks to the booby birds.

Needless to say, my welcome home from the war was great and I was treated like a prince by the girls. I made the acquaintance of a younger brother, Charles, and a much younger sister, Libba. After a pleasant stay of a month or so, I should have returned to Turks where I held a very minor post in the Customs which had contributed a modest half-pay during the war.

There was only the occasional opportunity of transport from Jamaica to Turks at that time. I was offered a temporary job in Kingston Customs, and took it to pay my way while waiting. Late in 1921, my boss, the Collector of Customs, told me privately that he would like to promote me over the heads of several of his staff — all colored — but dared not because the elected members of the Legislative Council — also colored — would raise too much hell! Obviously, there was no future for me in that atmosphere and, besides, I was already thinking of new horizons.

Earlier in the year, I met two young men who had left the island to work on the banana plantations of the United Fruit Company in Honduras. I was immediately interested in their tales of life there, a life of real appeal to a single man in search of new scenes in far-away places. In my job in the Customs I had recently returned to the drudgery of office work — the inspection of invoices from nine to four — after three months on my own as boarding officer of ships arriving at Port Royal, the entrance of Kingston's magnificent harbor. I had enjoyed that job, its occasional perquisites — gifts from friendly captains and chief stewards, a launch and a sailing whaler at my disposal, and the hospitality of the Royal Engineers Officers Mess, which was shortly to close after a hundred years of active life. The Customs officer I relieved returned from sick leave, and I went back to the inspection of invoices in Kingston. It was then that the Collector so clearly predicted my future if I were to remain in government service, which made it easy for me to decide to apply for the job on the banana plantations in Honduras.

The United Fruit Company was then in the process of expansion due to the demand for bananas and the ravages of the Panama disease, a soil fungus which proliferated rapidly, causing many thousands of acres to be abandoned. My application was readily accepted by the UFC manager in Jamaica and in February, 1922, I set off for Honduras in company with an acquaintance returning from vacation.

Just before my departure to Honduras, I spent a happy Christmas of 1921 with the family in Cayman. All of us were there — six daughters and two sons — and my good friend, Carl Nunes, who became my brother-in-law when he married Aline. My cousin, Franklin Harriot, came from Turks Islands to spend Christmas with us. It was the last time we were all together at home for sister Carolie died in an automobile accident in Mississippi in 1926. I did not return to Cayman until 1928 when I married Myrtle. Franklin, the companion of my youth, I saw only once again for a short hour in 1938 when our ship, returning from England, stopped briefly in Turks Islands. He died in Vancouver in 1958.

The return to Jamaica was again by schooner, but without the enjoyment of the first voyage on the *Fulmar.* Soon after leaving Cayman we were becalmed for two days in heat and discomfort. Finally, the captain took pity on the passengers and landed us in Savanna La Mar, a port near the west end of Jamaica. There I renewed acquaintance with Eric Clark who had returned to Jamaica with me on the S.S. *Bayano* in 1920. His father was a sugar planter and for some days I enjoyed their traditional and very pleasant hospitality, and then finally took a train to Kingston.

Life in Honduras

In February, 1922, I arrived in Tela, Honduras, to begin employment as timekeeper in the agricultural department of the Tela Railroad Company, a subsidiary of the United Fruit Company of Boston, Massachusetts. The first person I saw on landing was Bill Mais, an early acquaintance in Jamaica and now chief clerk in the head office of my department.

Twenty overseers and timekeepers supervised the farms at that time. The two largest farms each carried an overseer, two timekeepers, and two foremen. And there were semi-abandoned farms such as Nevada, where I worked after my first promotion from timekeeper, handling it alone with occasional help from new arrivals — an American, one from Britain, and a Nicaraguan. Such was the mixture of nationalities as, of the twenty-one employees present when I joined the district, I recall only five Latins and one lone Honduran. In those early days, the coast, where the bananas were grown, was "the place where people died of malaria," a place of last resort for Hondurans and Guatemalans seeking work. One cannot blame them as the highlands were not only free from malaria, but offered a climate often described as "Eternal Spring," besides offering some social life which the coast entirely lacked.

In the decade before my time, when the UFC started a great expansion of banana production, the north coast of Honduras, and the triangle of Guatemala bordering the Caribbean, were largely uninhabited. Around 1908, the completion of the railway from Puerto Barrios to Guatemala City promoted the planting of bananas in the valley of the Motagua River. There were no railways into the valleys of the Colorado and Ulua Rivers in Honduras. The well-drained valleys were covered in virgin forest and favored with an annual rainfall of eighty inches or more, which proved to be excellent conditions for growing Gros Michel bananas.

During this decade, labor was non-existent in that region. There were many scattered settlements of Caribs, a misnomer as they were full-blooded Negroes who spoke a dialect of their own of mixed Spanish and African origin, but they lived only on the beaches and refused to work inland. Largely self-sufficient, they were fishermen who subsisted on fish, cassava, plantains, and corn from their cultivations within walking distance of the beach. They were the seamen of the area, venturing forth in their dugout canoes from Belize to Puerto Trujillo, an old Spanish settlement in Honduras, a distance of some hundreds of miles. However, the Caribs never went out of sight of land for more than a few hours.

For these reasons, the Company had to import labor to fell the forests and cultivate bananas, to build railroads and houses. They came mostly from Jamaica where the Negroes already understood the essentials of growing bananas, and were ready to move to another country for higher wages. Here I am reminded of a short story of those early times which was told to me by Charley Dodd, a real oldtimer, who had worked on the construction of the railroad from Puerto Barrios around 1908. I met him in 1933 when I was transferred to Guatemala, where Charley was train dispatcher for the Company branch railway line.

"Things was rugged in them days," he told me, "but you could get a gal for a fish hook, and fish hooks was two for a nickel in the commissary." The story was better when he told it as he used a more primitive word instead of "gal."

From the district office I moved in a day or so to Jilamo Farm, one of the best in the district and which, in my time, produced the biggest bunch of bananas I saw in thirty-four years in Central America. It was nineteen complete hands and must have weighed one hundred and forty pounds or more. In banana growing, of the Gros Michel variety, a twelve-hand bunch is big and weighs more than eighty pounds. Fourteen hands is not uncommon and sixteen hands rather rare. A complete nineteen-hand bunch — one in millions!

Jilamo carried two foremen: one Jamaican and one Honduran. Foremen, like sergeants in the army, really ran the business. A good foreman was quickly recognized and valued beyond price. I don't recall their names, but no doubt they were of great help to me in the first months I rode with one of them, at the discretion of the overseer, who usually rode by himself over the whole area. The Jamaican was very helpful as he, of course, spoke English and could explain things in detail. Later that year, when I became assistant overseer with a twenty-five dollar raise over my former seventy-five (promotion was that fast), I met Cristino Galeas, the foreman of El Tigre, part of the bigger farm El Retiro. El Tigre was isolated, across the river from El Retiro, and had its own railroad spur from the main line. Because of its isolation, it was the farm that the "killers" moved to — those who had recently killed someone. Cristino handled them to perfection for he was quite a man, and not in the least afraid of them and they knew it. I am happy to add that, during my farewell visit to Honduras in 1956, I went to see him. He was retired, of course, and living in the small town of Progreso on a pension of perhaps forty dollars, of which, believe it or not, the IRS deducted 30% of the Company's contribution, or about $6.50! I always thought that the Company did not fight hard enough against that iniquitous tax which was deducted in Boston.

I took to farm life immediately. It was interesting and took place out in the open, and I soon learned that increased responsibility was granted as soon as the individual showed sufficient knowledge of Spanish to be able to make himself understood. In my rides about the farm there were always interesting happenings to attract my attention, birds I had not seen before, fish, if the river was clear, and the occasional alligator in the *lagunas* of the Jilamo River. There were monkeys in the forested boundaries of the farm and frequently bands of *pisotis,* a near relative of the raccoon, which is as active in climbing as monkeys. They are very inquisitive and not easily frightened if one keeps quiet and doesn't move.

A picture comes to mind. I was out early, just before sunrise, probably to carry the *aviso* to a distant labor camp, i.e., the order to start harvesting fruit. In my early ride I passed by a *laguna* of the river, a curve which had been flooded and dammed and never reopened, thus it provided a narrow lake often full of fish. A number of Hondurans were in the water, the women bare-breasted when bathing or washing clothes. Some were fishing but using only the primitive methods of the Indians from hundreds of years before the arrival of the Spaniards. They used the stem of the vine, called *barbasco,* which they bruised or crushed in a small dugout canoe and then scattered over an area of the *laguna.* The toxic qualities of the vine quickly dispersed and fish rose to the surface.

There was something primeval in the scene. Wisps of fog were still rising from the *laguna,* the topless women, the tall trees in the surrounding jungle full of screeching parrots — all composing a scene similar to a fishing party of Indians many years before the Spanish conquest.

Ours was a rugged life, a way of life which no longer exists. There was only one married couple in the district, Uncle Bud Faust and his wife. Americans from Louisiana, they lived at district headquarters. Uncle Bud was older than average, hence the nickname. I believe he saved my life, as you shall hear.

All of us who worked on the farms carried guns openly, as did some of the laborers, for a Honduran's first ambition was to buy a gun as soon as he could. Next came a five-gallon hat and finally a horse, when he became a contractor and earned more money. The Honduran laborer was tough and it was said that his ancestors gave the most trouble of all to the Indian races which opposed the Spaniards. He was quick to take offense if the *Americano,* who usually spoke poor Spanish, unknowingly offended him or failed to pay the promised wage, but he was a hard worker and honest. We left our houses for weekends and never locked doors, even if there were keys to fit them, which was unusual.

Paydays were more often than not enlivened by drunken fights and, since the laborer was never parted from his *machete,* a twenty-six inch, broad-bladed knife which was kept as sharp as the soft steel allowed, there were frequent killings and serious wounds. The winner had only to cross the river to escape the perfunctory search of the *comandante* and his few soldiers. The Company sold no liquor in the farm commissaries but did sell Bay Rum which many drank when the local, illicit stills were out of a brew called *aguardiente,* probably the first run of spirits in the very crude distillation from molasses, which smelt abominable and probably tasted worse. They drank this or Bay Rum, followed by a chaser of local beer which tested at 12% alcohol.

An incident in which I was concerned depicts very well the character of the Honduran laborer. Born in what we would call poverty, accustomed to a sparse diet of corn, black beans, and cassava, only occasionally flavored by beef or chicken, he was a tough man. As the saying goes, "as independent as a hog on ice!"

We occasionally had trouble with squatters and it was the overseer's duty to report all felling of woodland on Company property. In such cases, the Honduran was only following the age-old practice of his ancestors, that is, felling a few acres of forest in the dry season to be able to plant his *milpa,* or corn patch. If this was protected from deer, coons, and parrots, it provided the staple food of his family and pigs. He, therefore, saw no reason why he could not continue with this practice in the extensive forests of the coast, which was probably part of grants to those who followed the *conquistadores* and whose heirs had never visited them.

I never could understand why the Company made such a fuss about squatters in our district where all available banana land had already been planted, but the management considered it important to make squatters move, even paying for their labor involved. On the Nevada Farm, one morning I heard the sound of axes and falling trees. I went into the forest with my foreman to investigate and soon found the intruder, who admitted that he was on Company land but that he had felled only a small patch to plant his *milpa*, but, no, he wouldn't accept money for his work and move. I reported this to the district superintendent, Mr. Chute, who merely replied that we would have to get him off the property.

On his next inspection of my farm, he asked if I had induced the squatter to move off.

"No," I replied, "he's still there and has burned off his patch and planted corn. He seems really tough," I added.

"Hell's fire!" said Mr. Chute, "You must keep after him and make him move. I want to talk to him myself this morning."

We mounted our mules and rode off to various parts of the farm where work was in progress. After awhile I said, "We turn off here into the forest if you want to talk to the squatter."

A short ride on the well-marked trail brought us to the shack which the man had built of convenient timber and palm-thatch. Perhaps his mongrel dogs warned him of our coming, for, as we rode up to his shack, he was outside, sharpening his machete with long, soft strokes of the file. We said "*buenos dias,*" politely and were politely answered back. In ordinary tones, Mr. Chute reminded him that he had no right to fell woodland on Company property. Finally, he added that, if necessary, we would bring soldiers to move him on. That did it! The squatter answered the threat in the hardest tones I had ever heard, "*Esta es mi casa!*" — This is my home. "*Soy Hondurano y soy hijo del pais, y ni ustedes ni niun otro hijo de puta van a quitarme de aqui!*" (I am Honduran and native of this country and neither you nor any other s.o.b. is going to run me off from here!") There was complete silence except for the soft strokes of the steel he continued to sharpen — an already sharp machete. He meant what he said and Mr. Chute knew it. "Let's go, Hutchings," he said, and we rode off. The squatter stayed as far as I remember for I did not go back to find out.

I suppose I stayed on Jilamo about six months, then was promoted to assistant overseer of El Retiro, in charge of El Tigre, perhaps about three to four hundred acres across the river. I was not happy to leave Jilamo and Bill Vernon, and was doubtful about living with the only Honduran overseer in the district, whose name I don't recall. But these doubts were easily removed by an increase in salary to one hundred dollars a month, already a bit more than the salary I had received in Jamaica. It was an early promotion as, in those days, it was said that the Company needed two additional men for every timekeeper employed — one going, one coming, and one on the job — but I had a good ear for Spanish even though I studied it very little. Indeed, when my Nicaraguan timekeeper made me a present of a textbook entitled *Los Diez Mil Verbos Españoles,* I thanked him but said, "If there are ten thousand, then the hell with them!" I did not have the sense to ask him to speak only Spanish to me — more's the pity.

There was little change in the work on El Tigre Farm. I liked the necessary fording of the river mounted on a mule, although occasionally I had to wait for an hour or two to get across without swimming. Cristino Galeas was my foreman there and I quickly appreciated his worth, and had to speak Spanish to him. Unfortunately, the Honduran overseer had lived in the States and spoke fluent English, but I did get to hear more Spanish than in Jilamo.

There was no recreation except hunting which did not interest me as it was only at night, with a spotlight to shine the eyes of the game. Deer and *tepescuintle* — the latter a large forest rodent which made excellent eating — were plentiful. Most animals were nocturnal in habit, even the deer, as they holed up in the bush during the day and could not be found. On Sundays we sometimes went to the district headquarters to drink too much whiskey. Or once a month, on Saturday afternoons, we would take the train to Tela to stay over until Sunday afternoon. Tela was the port and Company headquarters, and bachelor quarters were maintained with spare rooms for the "visiting" farmers. One such trip with Bill Vernon and Uncle Bud reminds me of a time when the latter acted as my guardian angel.

There was little recreation available for bachelors in the Company town — two pool tables and that was all. The "farmers" usually crossed the river to Old Tela, which had existed long before the Company arrived. The meeting place for all the "farmers," newcomers, drifters, etc., was Dave Balderac's bar and rooming house. Dave was a Belgian and the most generous of hotel-keepers. According to his reputation, he had never turned down a white man who asked for room and board, though I doubt if there was anyone more experienced in recognizing those who were drifters and not likely to be in funds. Consequently, he had a surprising collection of valuable and near worthless articles accepted over the years from drifters who could not pay. He told me he had sixteen watches and other items of jewelry. His collection included jaguar skins, Indian bows, and four Jivaro shrunken heads.

Bill Vernon, Uncle Bud, and I spent several hours in Dave's establishment one night, which included a small dance hall. Bill did not drink much and Uncle Bud less than that, but I had taken on more than I could carry.

I guess it was after midnight when we decided to return to the bachelor quarters. On the way we thought it a good idea to have a bite to eat in the Company's all-night snack bar, which was operated for the benefit of the railroad and wharf employees on night duty. The privilege was extended to the government customs officers when there was a ship tied up to the wharf. When we arrived, there was one lone customer, a Honduran, standing at the bar. I ordered ham and eggs. We were all standing at the bar and I was next to the Honduran, not too close to him, with the others on my other side.

After a while, the cook brought a plate of ham and eggs and placed it on the counter between the Honduran and myself. It must have been closer to me for I took it, not knowing that the Honduran had ordered ham and eggs before we came in. He snatched it from my hands and plastered it all over my shirt front, saying nothing when a few words would have invited my explanation. To him, I was an instance of American arrogance which he justifiably resented. In my state of near intoxication, I could not take the insult, deliberate and uncalled for, and said nothing either, but went for him. The Honduran had backed away a bit and, before I could reach him — thank God — Uncle Bud jumped between us with his hands on my chest and his back to the other.

"You goddam fool!" he shouted. "Don't you see he has his gun on you?"

He had it on me, held waist-high and pointed at my belly. If I had reached him and hit him, there is no doubt at all that he would have shot me. Of all the Latin races I know, only the Costa Rican is inclined to fight with his fists, and the rest invariably resorting to guns, or machetes if within reach.

I was almost sober within minutes after the incident. The Honduran moved out, having sufficiently demonstrated his *machismo.* I don't recall that we discussed the incident as it was too clear what might have happened. I hope I thanked Uncle Bud adequately — I know I did with heartfelt thanks the next morning. Shortly after, he returned to the States and I never saw him again.

In those days, 1922-25, the Honduran farm laborer was tough to handle, especially for those who did not speak fluent Spanish and were strangers to Latin customs and thinking. Nearly all our labor came from the interior, some as far as the borders of Nicaragua which was perhaps one week away on foot. They really had to be tough because, on their return home, they had to outwit the bandits who readily guessed that anyone coming from the coast usually carried several months' earnings. In that period, only the Jamaicans remained on the farms throughout the year. All others were from the interior and worked only long enough to hoard up a small stake before returning to their homes.

Thus, the Honduran laborer hardly paused in his jump from mule paths and ox-carts to the airplane. In the late twenties, Yerricks, a New Zealander and pilot in the First World War, started the TACA airline with two six or eight-seater planes — Bellancas, I believe. It was an early success as TACA quickly established inexpensive airstrips where needed and quite literally carried anything, in addition to passengers. Our laborers realized the advantages of air travel immediately — a few hours, including the indefinite wait at the airport, compared with days on foot and risk to their earnings. Yerricks eventually sold out to bigger intersts after he lost an eye flying a Honduran general during a revolution.

Under the prevailing conditions, the handling of tough, itinerant labor in the complete absence of police and respect for the law — and, above all, little or no knowledge of the habits, customs, and thinking of the laborers we had to supervise — there were occasional casualties. I mishandled one situation in ignorance of the other man's thinking which led to serious consequences. I shall detail it later.

Folger was a new arrival, about 1925-26, I think. I remember him mainly because he showed me a clipping from his hometown newspaper which mentioned his early departure for Honduras, saying that "He would be missed as he was the best pecan-budder in the county," in Louisiana, I believe. A quiet, decent fellow, he should have returned to his pecan-budding after the novelty of Honduras had worn off. I don't believe he was with us more than a year and some months and, while acting overseer, he had the unfortunate happening that most of the farm mules got out of the pasture one night and gathered on the railroad track. A locomotive hauling a long train of fruit cars rounded a nearby curve and ploughed right through them killing fourteen, as I recall it. It may be that the superintendent — thank God, not I — bawled him out, but Folger went to the labor camp around noon, bawled out his stockman and fired him, all in very poor Spanish, and in the presence of the other laborers. The stockman had a grievance in that he lost face in being reprimanded in the presence of others, also he claimed that he had left the pasture gates firmly closed, which was quite likely. Some mules are clever at getting out of fences and many will follow the first one out. Folger was not armed at the time, but the stockman carried a gun and killed him. It was perhaps the saddest incident I know of as Folger was a very quiet, inoffensive type, but his inexperience led to his death.

There is one more incident worth telling of the violent reactions common in the life of the Honduran laborer of the times. More often than not, he could not read or write and, what was not so common in Latin countries, there had probably been little religious influence in his upbringing.

The United Fruit Company operated two divisions in Honduras during the period I am writing about: one from the port of Tela and the other from Puerto Castilla of the Truxillo Railroad Company, about a hundred miles east along the coast. Both ports were Company-built and, if at all possible and convenient in steamship schedules, it became customary for one division to visit the other for the big Fourth of July celebration. This was always celebrated with a baseball game and a dance, with side-lines of big dice games for the gamblers.

The rivalry in baseball was high, and several thousand dollars changed hands with the results of the game. The farms supplied most of the players. Weekend practices started a month before the game and, during the week of the big event, all players lived in Tela in order to practice every day. As often as possible during that week, two teams were made up to play a realistic game.

It happened that, during one of those practice games, the crowd of Honduran onlookers was more than usually difficult to keep off the foul lines. Finally, four players, all with baseball bats held laterally across their chests, endeavored to hold the crowd back behind the foul lines, quite casually and without roughness whatsoever. They were successful at first as the front members of the crowd understood the reason for it.

I knew one of the Americans who was employed in this task, Jack Zwicker, a good-natured, pleasant chap, and powerfully built. Perhaps in some exciting moment of play, the crowd surged across the foul lines and Jack tried to move them back, pressing laterally with his bat as described. A man in front rank reached out with a knife and stabbed him in the belly. Jack swung the bat and hit him so hard on the side of the head that the other side bulged out. Very fortunately, no disorder followed as very few actually saw the incident, all eyes being on the game. Our Dr. Nutter, well-known and liked by many Hondurans present, insisted that both men be taken to the hospital immediately. The game stopped and the crowd dispersed peaceably. The Honduran never regained consciousness and died that night, while Jack was not seriously wounded, no intestines having been pierced. He recoverd quickly but remained in serious condition as far as the local authorities were concerned, who had placed only token guards at the hospital to prevent his escape.

The Tela Hospital was almost on the beach, only a few yards from the water. One night, not long after the incident, one of the Carib dugouts, which I mentioned before, slid up on to the beach in front of the hospital with clearance for Belize, British Honduras. Jack was ready and, after two or three days of very uncomfortable sailing, he reached Belize, where the Company agent took care of his return to the States. No doubt, the Tela *comandante* was taken care of, he having no fear of adverse reports being sent to Teguigalpa, the capitol, as it was as far away as in the days of Spanish rule, many days by muleback.

Yet another incident comes to mind which has some relation to the violence in the character of the Honduran. The small, weekly paper of San Pedro Sula carried the story of an *hacendado,* or landowner, who was riding into San Pedro Sula from his hacienda. Picture him in his big hat and embroidered jacket, pistol in holster, and riding a gaited stallion (mares were not considered riding animals). As he entered town, he met a peon unloading firewood from an ox-cart drawn across the narrow, cobbled street, leaving no room for the "big shot" to ride past. No doubt, politely, he asked the peon to move the oxen a bit so that he could pass. The peon, equally polite, replied, "*Si señor,* but wait a minute until I finish unloading." This was too much for the *hacendado* and, hand on gun, he said, "*Pero ahorita*" — right now. Perhaps there was some discussion of the virtues of their respective mothers before the *hacendado* killed the peon. The paper made no mention of arrest.

A few days later, I happened to discuss the newspaper item with a Honduran who spoke fluent English. Probably my expression induced him to say, "But you don't understand how such a thing could happen, do you? He had to kill him, *por su honradez.*" I imagine *honradez* can mean many things, and in the incident mentioned I guess it meant "self-respect."

But casualties were relatively few among overseer and timekeepers during my time in Honduras. By the time I was made overseer of Nevada Farm, with a satisfactory increase in salary, I had learned enough Spanish, as I had a good ear, and accent was more important than grammar in conversations with poorly educated laborers.

Agricultural practices were simple in growing bananas in those times, as the Company had planted the naturally drained areas close to the rivers. Irrigation was not necessary in the short dry seasons which prevailed on the coast. I had plenty to do to fill my time for my supervision included the remnants of the Arizona Farm which was an hour's ride from mine. Most of Arizona had been abandoned due to the spread of Panama disease.

Nevada included an area of woodland which had remained intact except for the occasional squatter. In this sector, I spent many interesting hours with my foreman, Lisandro Diaz, who owned three *tepescuintle* dogs, mongrels no bigger than fox terriers but, with their near-starving condition, they were more than eager to hunt since, if successful, they were rewarded with the inedible parts of a *tepescuintle.*

The forest itself was interesting. I was first impressed with the silence, after the dawn chorus of parrots and howler monkeys, and I walked for what seemed to be hours, hearing nothing but the occasional twitter of a small bird, or the mournful call of some other distant bird which I never identified. I was ready to believe that the silence was due to the absence of animals until Lisandro explained that only monkeys and pisotis moved around in the daytime, and that all others were nocturnal in their habits. I had no conception of the vast numbers of other inhabitants until the first flood, when the river rose over its banks and flooded the forest to a depth of perhaps three feet. That evening, as it became dark, the voices of insects, frogs, toads, and certain birds swelled to a roar which reminded me of surf on a not-too-distant beach and continued unbated throughout the night.

During that flood, parts of the railway remained above water, but the bananas were flooded. Lisandro suggested we walk along the railway to see if any deer or *tepescuintle* had taken refuge in the higher areas. We found nothing until I heard a small animal crying not far away in the bananas. I moved into the water, more than knee-deep, until I saw a kitten, quite large, but still a kitten. Its features and tawny color convinced me it was a puma cub. It was perched on a log above the water. I wanted that kitten but, as soon as it saw me, it changed the tempo and meaning of its call and I heard but could not see, a much larger animal plunging toward it through the water. I carried a Webley pistol but decided I did not want to fight mother puma for her kitten!

Nevada farm produced a much rarer botanical curiosity for me than the nineteen-hand bunch of bananas I mentioned earlier, for which the odds were one in millions. Throughout the farm, there were many small patches of bananas grown up in bush, and abandoned because of Panama disease. Riding with Lisandro one day, he mentioned that a large boa constrictor had been seen to enter the patch the day before.

"Let's go see if he is still here," I urged.

We did not find the boa but did find its hole, a burrow about eight inches in diameter, worn shiny smooth by its passage. Probably a *tepescuintle* burrow taken over by the snake after he had eaten the occupant.

But we also found a far greater curiosity — a banana plant bearing two bunches of fruit, quite small bunches, as I recall, a five and six-hand bunches. The banana plant grows from an eye on a bulb. The eye sprouts a single stalk above the ground which develops a crown of leaves and, eventually, a bunch of the fruit. In the case I am recording, the single stalk divided into two about four feet above the ground, each with its crown of leaves and bunches of fruit, a previously unheard-of curiosity in which the odds of reoccurrence must be astronomical. There were many who came to see my freak banana plant in the few months of its life.

Vacation in Utila

In the latter part of 1923, I was due for my first vacation. I did not wish to spend it in Jamaica, where I had so recently lived, and I had no money for an expensive trip. I had heard elsewhere that the Company launch, *Victor,* was to leave for Utila in the Bay Islands, to escape possible seizure and use in the revolution then in progress, so I decided to go there for my vacation. It was a good choice for, not only did I enjoy the trip and a quite inexpensive vacation, but I also learned from a most reliable source, of the "tragedy" of the schooner *Olympia,* a tale of rape and murder on the high seas, and of a brave woman who saved herself by her own efforts. So it was that one afternoon I went by train from Nevada to Tela, the Company port and headquarters, and boarded the launch for the night run to Utila, perhaps forty or fifty miles distant.

The Bay Islands in the Gulf of Honduras comprise numerous small cays and three islands, of which only Roatan is of fair size. Bonacca and Utila are also inhabited but, of the three, Utila is the best known and with the most inhabitants. Their history is interesting to those familiar with the Caribbean. The islands were settled early in the eighteenth century by "turtlers" from Grand Cayman. Many of the inhabitants are of Anglo origin and still speak English.

We arrived in the early morning and, in recognition of the Company launch and its island captain and crew, all landing restrictions were waived. I went ashore immediately and engaged a room in the only boarding-house, run by a Mr. Bodden, whose surname is sure proof of Caymanian ancestry. Mr. Bodden was the father of sixteen-year-old Alicia, a passenger on the *Olympia,* who was murdered under terrible circumstances. It was he who told me of the tragedy and the events which followed.

The first few days in Utila passed pleasantly enough — as pleasantly as the sandflies permitted, that is! I should mention that part of the attraction of Utila was due to Nadine Hunter, a girl of about my age, whom I had met in Grand Cayman two years before and who was presently visiting her grandmother. The attraction was more than anything else due to her romantic life, accompanying her father, Captain and part-owner of a three-masted schooner which sailed between southern ports of the U.S. and various destinations in the Caribbean. So, the days were pleasant enough with Nadine to keep me company. There were other girls, some prettier, but few had journeyed beyond Honduras or had heard of the far ports and islands which Nadine knew so well. Theirs was only the quiet life of the islands, of little interest to me.

The only interesting incident of those first days was a fishing trip made with two young men to the west end of the island, which was mainly mangrove swamp and uninhabited. We sailed in a small, decked sloop, towing a dugout, and provisioned for three or four days. We slept on the sloop and there I heard for the first time the distant bellowing of alligators and was told that they were plentiful in the swamp. It was a very effective background in the night, while a recluse we met there told us stories of buried treasure. He almost induced me to come back and help him look for it!

We used the dugout for fishing and I remember one trip vividly. We were bottom fishing with hand lines and had caught quite a few fish when a shark appeared and took several fish before we could boat them. Cursing the shark did no good so we decided to move. The islanders in those days used a handmade cordage of thatch palm leaves in all small boats where no great strain was expected. It was cheap but with a very short life. We had anchored the dugout to a heavy stone carried by a "thatch" rope, and one of the fellows stood up in the bows and started to haul up the anchor. The "rope" broke and he fell backwards, upsetting the dugout. A heavy craft, it has two remarkable properties — it cannot sink and, if upset like ours, the experienced users can "rock out" enough water to enable one to get aboard promptly and bail out the remainder.

It sounded ludicrous when we talked of the incident afterwards. All three of us were bobbing our heads under water to locate the shark. In the upset we had lost most of the fish, some still alive. We soon saw the shark nearby, picking up the floating fish and paying not the least attention to us! After one failure when someone got into the dugout too soon and it filled with water again, we managed to rock it out and bail it dry, but our misfortunes were not over for the day. We cooked all meals ashore and that morning had left a big, covered pot slowly cooking a *caldo* of crayfish, plantains, and cassava. We found the pot bare of all but the remnants of these savory ingredients and noted a bunch of ever-watchful *zopilotes,* or buzzards, hanging around. We deduced that they had overturned the pot as soon as the fire died down and eaten the contents. We really cursed the *zopilotes* as we had eaten nothing since early morning and had to wait for the preparation of another meal.

Soon after our return from the fishing trip, one morning Mr. Bodden told me the story of "The Tragedy of the Schooner

Olympia". I am ashamed to say I cannot recall the name of the sole survivor whom I met a few days later. Many years have gone by since the story was told and I took no notes, but it was told to me by the father of Alicia, the young girl who died in terrible circumstances, and further verified by an interview with the lone survivor and heroine of the story — for heroine she was — a brave young woman with a stout heart and will to live while facing extreme odds. She is called Mary in this narrtive. I will call the villain Thomas.

Mary was a short, spare woman with plain features, in her thirties, I judged. I found her shy and reticent in conversation and perhaps there was some subconscious block in her mind when reminded of her desperate ordeal, for she answered most of my questions in monosyllables. I got no impression of the exceptional mind that sustained her thinking during the fearful events, of the will and character which triumphed over the physical exhaustion and drove her on to safety. But I am getting ahead of my story. I hope I leave you with the impression that I have clearly recalled the story of these extra-ordinary events which I now revive for you. My familiarity with the sea and sailing ships will enable me to convey something of the *ambiente* but no attempt will be made at dialogue.

The Tragedy of the *Olympia*

The *Olympia* was a two-masted schooner of less than a hundred tons. Still another Bodden was Captain and part-owner, and he sailed her to trade along the north coast of Honduras, east and west from the port of La Ceiba. He supplied many small and primitive settlements along the coast with the essentials of flour and salt, machetes, axes and files, clothing, cotton piece goods, etc. In return, he took sarsaparilla roots, a few grains of alluvial gold washed from the creeks by those who were satisfied to earn a dollar a day, but his main business was the purchase of cattle and hogs, and transporting them to La Ceiba for butchering. He paid for the animals in dollars and was known to carry money, up to a thousand dollars or so. A Negro seaman, Thomas, who lived in Utila and had made several trips in the *Olympia*, knew of the dollars usually carried in her strong-box and made his plans to get them. Simple plans, for he needed only the opportunity. In this plan, he already had a perfect alibi as protection against any suspicion, as will be seen later.

It was the custom in the islands, on the departure of a vessel for the mainland — in this case to La Ceiba — to sail in the early evening in order to arrive the following morning round eight or nine o'clock, at which time a customs official could be expected to board her. A slow passage over approximately forty miles, but the sailing vessel would be beating dead to windward against a light land breeze. If there were passengers, it was customary to bring the schooner in the late afternoon as close to the shore as was safe during the few hours' wait, in which case the crew would be allowed off to see their wives and sweethearts, collect their clean clothes, and eat a last meal at home. Thus, the schooner would have no one aboard from sunset until perhaps eight in the evening when the passengers arrived.

Having sailed several times in the *Olympia*, Thomas knew of these last minute arrangements and counted on them. They were essential to his simple plans, which only required that he get on board the vessel unseen. The fact that he lived alone was vitally important as he would not be missed overnight which was all the time he would need if his plans went well. The only preparations he had to make were two small hardwood wedges no more than six inches long, but he knew where the axe and firewood were stored aboard. They were the only other essentials. As is the custom of seamen, they would certainly be in their usual place, in the corner of the hold and handy to the hatch. So, as it became dusk, he moved unobtrusively to the shore, as close to the schooner as possible, in a suitable hiding place. When night fell he was certain that no one had remained aboard. He stripped and, not forgetting his wedges and tying his shirt and pants on his head, he swam unobserved to the schooner and climbed aboard by the bowsprit stay. Creeping below the level of the gunwale, he moved immediately to the open hold and, finding the axe and firewood in the accustomed place, he sat down and built up the firewood around him to make a protective screen whose increased bulk would not be noticed in the dark. He knew that no firewood would be required, nor any fires lit, in the short night passage to La Ceiba. He waited there comfortably for the few short hours yet to run in the life of the *Olympia* and all but one of the passengers and crew who sailed in her that night.

The passengers probably came on board around eight o'clock. They were Bodden's wife and three week-old baby, Alicia Bodden, the sixteen year-old daughter of the boarding-house proprietor, and a woman of about twenty-five whom I have called Mary for the purposes of this narrative. All the adults had certainly made trips to La Ceiba before and were experienced also in traveling in small sailing vessels. As soon as their gear was stowed within reach and the vessel underway, they settled down in the cabin for the night.

An hour later the schooner had rounded the end of the small island and started the hourly series of tacks, or changes of course, as she zigzagged against the fresh land breeze blowing from the mainland. There was the steady rush of water against the bow, the creaking of the heavy blocks and the pattering of the reef points against the mainsail — all well-accustomed sounds, even to the passengers who were sleeping soundly.

There would only be two men on deck: one in the bows on lookout, but probably half-asleep as he would be roused only every hour or so when the schooner tacked and he had to tend the sheets of the jib and foresail, and the one at the wheel, his face dimly illuminated by the compass light. He would not be sleeping. The movement of the ship under his hands and the "shivering" of the mainsail peak if he sailed her too close to the wind, keeping him awake. The other two seamen were asleep in their bunks in the extension of the hold toward the bows. No bells had been struck, but probably around eleven o'clock Thomas judged that all those not on watch were asleep. The schooner was perhaps six to eight miles to windward of the island which lay directly down wind, as he had hoped. A seaman himself, he judged that the various tacks had been consistent in duration and within the extremes of the island.

Softly removing the firewood, the villain stood and crept quietly to the forecastle to assure himself that the two seamen off duty were asleep. Carrying the two wedges and the axe, he crept on deck. As expected, he saw only one man on lookout in the bow and one at the wheel. He then softly slid the hatch closed, and with the axe, he gently tapped the wedges into the grooves in which it moved, one wedge on each side. The two men sleeping below were then imprisoned and no power of theirs could release them, for a stout bulkhead partitioned the hold from the after-cabin.

He then boldly moved toward the lookout in the bow who, if he saw him, would only think he was one of the crew. We can hope that the lookout was drowsing and never knew what hit him — one blow of the axe, the dull impact and falling body overboard was lost in the many noises of a sailing ship at sea. He then moved aft to the man at the wheel — a different problem, for he would be seen and had to face his victim. Once more, moving casually, he aroused no suspicion until within reach, when he killed or stunned the man and threw him overboard. On the release of the wheel, the schooner surged abruptly into the wind, with all the noise of big sails flapping and the jolt of heavy booms swinging across the deck. The Captain was instantly roused and rushed on deck, but Thomas was probably waiting for him on one side of the hatch, and killed him also. Probably he then went immediately to the cabin to look for the Captain's pistol and found it in its accustomed place. In a few short minutes he had become master of the ship, the passengers, and the remaining crew.

Imagine, if you can, the terror of the mother when aware of her husband's death, of her approaching death, and that of her baby. Perhaps it was not yet equalled by the terror and anguish of Mary and Alicia whose imagination could not cope with fate which awaited them. And all of this while hearing the muffled cries and hammering of the men imprisoned below, the unusual crashing noises of heavy booms and tackle swinging across the deck, the schooner hove to, pitching and rolling in the short sea, and in the terrible confusion induced by being startled awake from deep sleep, to face unbelievable conditions of terror and death.

Thomas undoubtedly knew of the intended passengers before he boarded the schooner. Their presence could only excite his more vicious tendencies when his plans succeeded and, after the murder of the three men, his inhibitions, if any, were released. So, we shall not speculate on the added torture of the poor mother whom he had already decided was to go down with the ship. No lesser fate was reserved for Mary and Alicia, but not before he willed it.

With the useful axe, he tore out the floorboards of the cabin and began chopping a hole in the very bottom of the ship. She was to be sunk without a trace for, notwithstanding her wooden hull, her ballast would carry her down into the depths. The two imprisoned seamen probably felt and guessed what the axe blows meant and, with the first rush of water, their shouts were changed to screams of mercy, but there was none for them either.

The schooner carried a small dugout canoe on deck, capable

of carrying three or four persons. A heavy craft, but useful for landings in the surf of some anchorages on the coast. With the help of the two women, Thomas got the dugout overboard and forced them into it at pistol point. Taking the Captain's money bag, which I was told held approximately $1,200, he pushed off from the sinking ship. It was then about midnight, and probably six miles or more to the windward of Utila. With a following wind and sea, the two women to paddle, he would reach the windward shore before daylight, in ample time to enjoy the women — before murdering them.

By then, Mary had passed the stage of confused terror and had started to think again. Sometime after leaving the schooner, she dived into the sea and swam away from the dugout. Thomas shot at her several times and hit her once through the shoulder but without hitting the bone. She probably dived often and he would find it difficult to turn the heavy dugout, and she was able to swim away from him in the dark. He thought she was seriously wounded and could not survive the long swim to the island, so he left her and resumed his way to what he thought was safety. But Mary was an island girl and a strong swimmer, with a stout heart and the will to live under extreme adversity. Fortunately, she recognized the direction she should swim in — down wind in a following sea — fortunate indeed, as she did not have to swim against the sea which would have required much greater effort and the frequent slap of spray in her face would have greatly hindered her breathing. She swam slowly, mostly just drifting to conserve her strength while the following sea pushed her along, a friendly sea for nothing harmful neared her through the long hours of the night.

But what of Alicia? She must have wanted to follow Mary and prepared to dive overboard as well, but was instantly menaced with the pistol. Poor child — her immaturity could not provide the reasoning and strength of purpose that Mary showed. The menace of the gun paralyzed her and she lived a few more hours.

Like most islands in the Caribbean, the settlement of Utila is on the leeward side, away from the prevailing winds, with beaches and calm waters during the greater part of the year. The windward shore of Utila was known as the "Iron Shore", and was well named. There are no beaches, and the soft coral rock is eroded by wind and spray into honeycombed pinnacles, as sharp as razors — pinnacles which break under any weight and causing further wounds if one attempts to walk on them. A barefooted person could not walk a half mile, and the stoutest shoes could not last more than two or three miles. Since the greater part of the interior of Utila consists of mangrove swamp between the windward and leeward coasts, very few people went there, and only one path existed for the occasional fisherman to get to the "Iron Shore." Thomas, a sometime fisherman, knew of the path and the landmarks which located it from the sea. When daylight came, he approached within easy distance of the path. The villain then raped and murdered Alicia and sunk her body in deep water. He holed the dugout, loaded it with rocks and sunk it as well. He then took the path to the settlement, meeting no one on the way, and arrived before noon with the sure knowledge that nobody knew he had left the night before. We would not know of the "mills of God which grind exceedingly slow but grind exceedingly fine" — but exceedingly fast, in this case, for him.

We must return to Mary, perhaps six miles to the windward of Utila, swimming in the dark night with a bullet wound in her shoulder and still suffering from the shock and mental anguish of the past few hours. Certainly she prayed for, in those times, the church and religion were a powerful influence in the isolated life of the islanders. She had faith in prayer and her faith and courage sustained her — how else could she have survived the long hours of the night? She had also the islanders' fear of sharks, but they are not too prevalent in the deep where their food is scarce. The more serious danger would arise when she reached the reefs and shallows of the coast. What were her thoughts when she saw the sunrise above the lonely waters? The sight of land, low on the horizon, heightened her hope and courage and she now knew that she would live. Some hours after sunrise, she passed unscathed through the dangerous area of the reefs and staggered ashore through a cleft in the rocks of the "Iron Shore."

Her first thought was of water to quench the terrible thirst brought on by the frequent splashing of saltwater in her mouth during the night. She might have found some in the eroded hollows of the rocks if it had rained during the night, but she told me she found none. Mary undoubtedly knew of the path across the interior of the island to the town, but she did not know which way to look for it, nor how far away it lay. Barefoot and near naked — she had left her shoes in the dugout and stripped away the long, clinging dress, as soon as she could in the sea — she had little protection against the sun and insects. She immediately discarded the idea of following the coast line around the island to the town as, barefoot, she could not have walked a quarter mile across the "Iron Shore" and the scrubby jungle above it was almost as difficult. Again she chose the only practical alternative, to cross the width of the island through the mangrove swamp to the open beaches of the leeward side, where she would recognize the landmarks and know which way to turn towards the town. She knew it would be difficult to move through the mangroves, but she did not guess how very bad it would be but, for her, it was the only way, and the short way, she told herself.

She rested only a short while and, before noon, her remarkable courage and tenacity had pulled her to her feet. After crossing the narrow belt of scrubby jungle, the land sloped down to the swamp and she entered the mangroves.

The tropical mangrove grows well in the shallows of salt and brackish water, to the height of around twenty-five feet. It grows no recognizable trunk as is the case of trees in the temperate zones. The widely spreading branches are supported by a pyramid of strong roots extending from the mud to a height of about three feet above the water. The trees are closely packed together, with little or no room to move between or over the pyramids of roots, armed at the water's edge with numerous small shells of clams and other shellfish, even small oysters, all with razor sharp edges. Thus, a mangrove swamp presents an almost impenetrable barrier to all except those who must cross it as a last resort. This was the case with Mary, and she took it with the same courage that saved her the night before, but even that courage must have faltered if she had known how many hours it would take to cross the swamp to safety.

First, she waded through quite shallow water and mud not yet knee-deep, laboriously finding a way over or between the widely spaced roots. Her bare feet and legs soon suffered many small cuts from the shells that armed the roots. Her mouth and throat were parched from the salt overnight and the lack of drinking water. She had to guide herself by the sun to continue on a straight course; to get lost now was unthinkable. Her progress was painfully slow and became a nightmare — and worse when darkness found her still in the swamp and the mosquitoes came in swams. She had little, if any, sleep that night, probably some lapses into unconsciousness, her tired body sitting in the water and leaning against the mangrove roots. Perhaps a fallen tree provided some rest above the water if the mosquitoes could be endured. She waited for daylight, she longed for daylight, thinking she could no longer endure the torture of thirst and mosquitoes.

The morning after the return of Thomas, the second morning after the sailing of the *Olympia*, two men sailed around the end of the island to the windward shore where they had set turtle nets inside the reef. They found the Bodden baby in the nets and recognized it at once, realizing that the *Olympia* was lost. Carrying the baby, they hurried back to town to spread the news and prepare to search for possible survivors.

Daylight came at last for Mary, after what must have seemed an interminable night. The sun rose at last to renew her courage and direct her course. Resting frequently now and near to exhaustion, she struggled on in her desperate journey. In the early morning, she probably reached open water, too deep for mangroves but not too deep for her to wade, except in the channels scoured by the tides. She would be heartened by the thought that the point where the mangroves started again on the far side was perhaps not distant from the beaches she had to reach. She crossed the open water, perhaps as much as a mile of relatively easy going, then once again had to face the mangroves and the nightmare of crawling in mud and water through the closely packed roots, but she kept her wits and sense of direction and struggled on. She must have rested long and frequently for it was late in the afternoon when she came out into the clear of the leeward beaches. She recognized her location and, no doubt, with gratitude that her prayers had been answered. But her trials were not yet over. Just as dusk deepened, she saw two men walking the beach toward her and stumbled to meet them, crying and waving her arms, her hair all awry, sunburned, and spattered with mud. They recognized her in spite of her bedraggled condition, saw that she was Mary who had been lost on the *Olympia* and concluded that they were looking at her ghost, and fled terrified back to town where they spread the news.

Their story quickly reached Thomas who was not at all disconcerted by the finding of the baby that morning, for no suspicion had touched him. But he was the only one in town who knew that this was no ghost approaching — it could only be

Mary, the woman he shot and left to drown in the ocean some miles to the windward of Utila. His luck had not yet run out, as it happened there was only one sizeable sailing craft anchored off the town at that time. He swam off to her in the dark, hoisted sail, and got away.

Still alone, Mary stumbled into town for nobody would go to meet her ghost in the darkness. When she was revived enough to tell her story, preparations were made to go after Thomas but they had to wait a day or two until a schooner arrived and was made available. By that time, Thomas had reached the mainland, but his pursuers knew of the small settlement along the coast where he was born. Touching first at La Ceiba to inform the authorities of the loss of the *Olympia* and the murders, they went directly to the tiny settlement of Providencia and, when they asked for the villain by name and told of his crimes, the people of Providencia turned him in to pay for his misdeeds. They took him back to Utila where he made a full confession. They then put a small chain around his neck and hung him from a mango tree; the tree and chain were still there in 1923.

Progress in Progreso

As I recall it, I remained on Nevada Farm until January or February, 1924, when I was transferred to the Progreso District to work with the engineers putting in the first irrigation for bananas in Honduras. Three thousand acres formed this project, located on five farms at the end of the Company railroad in the valley of the Ulua River, about a hundred kilometers from port. A much drier climate prevailed there, the dry season usually extending from mid-January to the end of April, with no rain.

After a night in the Progreso headquarters, located quite near the small town of that name, I moved out to Farm 12, more or less the center of the irrigation project. Octavio Schofield was the farm overseer. I had met him before in the Colorado District and now, living in the same house, we formed an intimate friendship which lasted his lifetime — perhaps forty years. I could almost write a book about Tavo from the many stories he told me of his father and their home in British Honduras, and of Tavo's own experiences in collecting chicle, then the base of chewing gum, in a province of Mexico.

Tavo's father was an eccentric Englishman, but when he died, he willed seventy-five thousand dollars in cash to each of his eight sons and daughters, or their equivalent in land and business assets — eight fortunes in those days! He was a man so far ahead of his time that he insisted on planting three mahogany trees for every one felled. Also a recognized authority on Mayan culture and religious practices, he literally believed that he was the reincarnation of a Mayan high priest and celebrated the important festivals. Tavo's mother was Mexican from Merida, with an unusual education for a Mexican woman of those times.

Tavo himself was a gentleman in thought and deed. He was the best man I ever knew in the handling of Honduran laborers. He knew them intimately and they trusted him. I once saw him take a machete away from a laborer who was howling drunk.

"My God, Tavo," I said, "that was a stupid risk to take!"

"No," he replied, "I know him and he wouldn't hurt me."

"Perhaps not, but why take the chance?"

"He probably would have hurt someone else or gotten himself killed," he replied.

Tavo was my son Harry's godfather, and I wish Harry had known him as I did. He served in Mesopotomia in the First World War. When he died in 1967, I was asked to write something of his life and service with our Company. I was immediately reminded of the opening words of the document of the King's Commission to all officers in the armed forces, of which Tavo was one. It reads: "To our trusty and well-beloved Octavio Schofield." Trusty and well-beloved — Tavo was all that to everybody who knew him. There can be no finer epitaph.

Tavo was overseer of Farm 12 and Jack Bates of Farm 11, each with a timekeeper, so with myself there were five in a two-bedroom house. Jack decided to move a bed into the farm office below. The house was raised on posts eight feet above the ground, leaving ample room for an office and storage space for implements. I shall have more to tell of Jack's decision later. My work was with the Engineering Department. I was inspector of drainage ditches and had to supervise the work of some hundred and fifty men engaged in digging an extensive drainage system which complemented the irrigation.

The engineers staked the proposed drains and I had to see that they were dug to the required specifications. I had no problems with the men or the sub-contractors but Charley Lanz, the German contractor, who lived in Progreso town, two hours away on horseback, could not come out every day to supervise

Harry H. Hutchings, District Superintendent, with bananas from Farm 12, Progreso, Honduras.

(Photo: Unifruitco, February, 1930)

his contract. I turned in only the figures of completed ditches to the engineers and, one pay day, Charley Lanz found that he had much less money coming than he expected. Someone had told him that he was paid according to my figures, and perhaps he suspected some skullduggery by which I withheld some of his earnings. At any rate, he came galloping on his horse up to the office, gun in hand. As soon as he found me, he said, "Be you not a white man and I shoot you!" but quieted down when he heard the facts, and we were good friends again.

Usually a gentle fellow, Charley had a keen sense of humor and a fund of very funny stories concerning incidents in his life. A well-educatd man and, from his own account, the ne'er-do-well son of wealthy parents still living in Germany. Of Charley's Indian woman, Maria, I shall have more to tell. Charley's English was not too good, and one of his funny stories I have never forgotten. He had twin boys by a native woman and, when they were about seven years old, he took them to the hospital clinic where it was decided they should have a dose of castor oil — at home. Next day Charley appeared with the boys and the story that all of the oil had been wasted in fighting the boys who had refused to swallow it.

"I bet I can get it down them," said the Irish head nurse,

rather rashly as it happened, for the boys fought and screamed and vomited to the extent that she finally quit trying. A compromise was reached when the boys agreed to take the dose if Charley would also. A compromise which only Charley would have accepted. Said Charley, "So I take it and, by God, I shit like a goose!" A bit earthy perhaps, but wildly hilarious when told by Charley.

The Farm 12 house was built about eight feet above ground, providing two bedrooms, a bath, dining room, kitchen, and porch on the top floor. The ground floor was ample for a screened office and implement storage room. As mentioned before, Jack Bates elected to place his bed there, in a corner of the big office.

The farm was about one hundred kilometers from the port, at the end of the railroad, and five days on mule from Tegucigalpa, the capital. A ride which even the manager had to take occasionally before the event of air transportation. The terminal was, therefore, the center of traffic to and from the interior and always of great interest to us. An incident I now come to, started there. In the telling of it — many times since those early days — I have called it "the reluctant virgin and water hazard."

One afternoon, Jack Bates got into conversation with a girl who had gotten off the train from the port, or was going there, and talked her into staying overnight. Obviously she was not a virgin, or her mother or protector would not have let her out of their sight, but my story needs an intriguing title! After supper that night I wanted something to read and decided to borrow from Jack's stock of *Adventure* magazines. He had gone to bed early so I went below to his room. I was not surprised to find his door locked, but banged on it and heard Jack say, "Just a minute," before he opened it. As I entered, to my surprise he closed and locked the door again and put the key in his pocket. I then saw the girl standing in a corner with her back to us and that there was water flowing over the concrete floor.

"What the hell is going on, Jack?" I asked. "She is a little shy," he replied, "and won't come to bed so I put on the garden hose from outside to help her make up her mind." (There was sufficient space between the siding and the concrete floor to permit the flow of water.) "Just a little shy," he added, "but she'll change her mind or sleep on the wet floor." After a few compliments regarding his ingenious solution of the problem, I moved out. Apparently it was solved as Jack expected, as the next morning the gal was all smiles as she helped the cook prepare breakfast.

Of course, a dragline dug the main drainage ditches and it did not take long for one hundred and fifty men to dig the subsidiaries. The average per man was eight cubic yards a day. I tried it, but my back muscles were not sufficiently developed to swing a spadeful of wet dirt the required distance out and away from the ditch. I then moved from the Farm 12 house to the engineering camp proper and was given charge of a gang engaged in heating crude oil and applying it to the main irrigation ditch, the idea being to reduce seepage. The results were not nearly good enough and eventually a much smaller, concrete-lined ditch was installed.

They were a friendly, carefree lot in the engineering camp at El Sapote and I readily fitted in. Dinty Moore was chief engineer and lived in a shack apart from the main bachelor quarters, the latter having three bedrooms. Charley Mitchell was senior engineer, assisted by Almyr Bump, a recent graduate, and three others whose names I do not recall as they did not stay long. "Doc" Sayer, English, had obviously been around, well-traveled, but never had anything to say about his past. A very pleasant fellow in a drinking party, he did not last long either and soon drifted off to other parts. He is remembered only because of one incident. In the early hours of the morning he woke Bump, and others, by calling in some agitation, "Bumpy, Bumpy, old chap, will you come quickly and throw my bloody mattress in the river; the damned thing's on fire!" I write Bump occasionally and always start my letters, "Bumpy, old chap!" As far as we know, he and I are the only ones left of that carefree crowd of 1924-25.

The irrigation project was ready for the start of dry weather in January, 1925. I was made watermaster to control the orderly sequence of applying water over the entire project — an interesting job but requiring long hours as the pumps shut down only after a quite infrequent rain. Big Worthington pumps they were, if memory does not fail, and they lifted thirty thousand gallons of water per minute from the Ulua River.

I was at a loose end for a month or so after the end of the irrigation season, mainly employed in the preparation of ditches for the next season. I then met Henry Hogaboom, an American from Vicksburg, who became my friend for life. We married in the same year and our sons were born about the same time.

Henry lived at the Progreso headquarters, quite near the town of the same name. He was in charge of the treatment of Panama disease, a quite useless treatment which was soon dropped. I spent most weekends in Progreso and, indeed, was founder member of the Progreso Athletic Club, which lasted all of three weeks before Ed Cobb, District Superintendent, broke it up. Some bright fellow discovered that the Tela Masonic Club and Golf Club were allowed to buy Scotch by the case at wholesale rates, so it seemed to be a good idea to found a club of our own. We did so and things went merrily along for a week or so, having a drink or so in the evening rather than go over to town to one of the bars. One day we opened the bar to have a drink while the office closed for the noon meal. It was nice to have an appetizer, we agreed, but we did not stop with the appetizer — in fact, did not stop at all until the cook got mad and closed the messroom and the chief clerk passed out. He did not return to work that day, and the next morning Ed Cobb broke up the Progreso Athletic Club. We always thought he did it because he was a bit of an alcoholic and was afraid to join the club. Anyway, we found that buying Scotch by the case only increased the bar bills!

The Progreso office staff put on an occasional Saturday night Victrola dance, to which we invited the local girls of respectable families. They were always heavily chaperoned — by the fathers, more often than not. On one occasion, the girls sent a verbal ultimatum that they would not come if Charley Lanz's woman was invited — Maria, at best half-Indian who boasted of more than one protector before she acquired Charley. So, Lanz was diplomatically informed that he was invited but Maria was not. He replied airily that we knew he never danced, but would come just for a drink. Apparently Maria had other ideas.

On the night of the dance, Charley and Maria showed up late but dressed for the occasion. Somebody must have intercepted them and induced them to go upstairs to the bar for a drink. The second floor consisted of only bedrooms and baths, and certainly none of the girls dancing below would dream of going up there, and a confrontation seemed impossible. Charley carried a belted gun which, in line with local courtesy, he took off and placed on the bed in one of the rooms where the bar was set up. He and Maria had lots of company as, of course, all the non-dancers were there. It is possible that Maria proposed they join the dance, which he refused and which started an argument fired by many drinks. It ended when Maria snatched Charley's forty-five from the holster and shot into the floor. The floor was only one inch pine and formed the ceiling of the room below, where perhaps six couples were dancing and others watching. The bullet ricocheted from the concrete floor and tore off the shoe heel of a fellow called Price who was dancing. I was not there at the precise moment, but was told that Price's expression was so ludicrous as he raised one foot and looked at his shoe that everyone started laughing without really considering what might have happened. There was much hilarity and the incident only livened up the evening, for I don't recall any of the girls leaving because of it. I saw one of them home — closely conducted by her father.

Planting at Guaymas

In the later part of 1925, I was promoted to assistant superintendent of the Guaymas District where a big planting program was underway. Draining, planting, and felling the virgin forests always appealed to me. By the end of my career, I calculated that I had supervised the planting of more than thirty thousand acres of bananas.

But there were many expensive hazards in the banana business, the worst being Panama disease. Wind storms and floods were also devastating. Therefore, the Company had to have vast reserves of land to constantly plant. In the early 1940's, United Fruit owned one hundred twenty thousand acres of bananas in six different countries of the Caribbean. This amounted to sixty million bunches annually, which included purchased fruit from independent planters.

So it was that early in this century, United Fruit moved into Honduras, where for some years a limited supply of excellent fruit had been barged to the mouth of the Ulua River. With the installation of a deepwater wharf at Tela, on the north coast, and a railroad into the valleys of the Colorado (Lean) and Ulua Rivers, a great area of banana cultivation was made available and was still further augmented by the establishing of a second division in Honduras, at Puerto Castilla, also on the north coast. From the beginning of development in 1914 at Tela and Puerto Castilla to 1924, United Fruit had planted 88,000 acres of bananas. In time, the Company held some 400,000 acres of land.

At the beginning of the century, there was competition among several American groups for banana lands along the north coast of Honduras and across the border in Guatemala. The Vaccaro Brothers, later Standard Fruit, operated from La Ceiba. Sam Zemurray of the Cuyamel Fruit Company expanded his interests from Omoa westward into disputed Guatemalan territory where United Fruit was, and eastward to the Ulua River. Upon his retirement in 1929, he conceded to United Fruit Company. This meant that the Company now controlled most of the banana lands on the north coast of Honduras, plus large plantings along the Motagua River in Guatemala.

In the planting of Company-owned lands in Central America, the Company wisely accepted the planting practices already established by the natives of the country, by far the cheapest and the only way to plant bananas in forested areas. First, a survey group measured and staked the boundaries of the farm, often determined by natural limits, such as streams, and as much as was practical, relative to the railroad, for experience soon showed that bananas sufferd grave bruising when hauled more than twenty-four hundred feet by pack-mule. I am writing now of the early days — in recent years, greatly improved transportation and packing has eliminated bruising. Next, gangs of men, working on a piece-work basis, underbrushed 1,200 by 800-foot sections — in Honduras roughly twenty acres after deductions for roads and ditches, etc. In underbrushing, all bush and small trees less than three inches in diameter were cut to the ground. Then came the liners, skilled men who could do as good a job in the forest as a surveyor with instruments. Their job was to drive a stake at the exact spot where a banana seed should be planted. They started from an engineer's stake at the corner of the section and were expected to arrive at the far end, twelve hundred feet away, with an error of no more than a few inches, after planting eighty-four stakes on the way, fourteen feet apart, assuming the planting to be fourteen by fourteen foot square.

Next came the banana "seed," bulbs like lily bulbs but weighing eight pounds or more, which were hauled from the railroad by mulepack and deposited, one at each stake. Then came the planters who dug a hole for the seed about eighteen inches deep and covered it. Finally, all standing trees were felled on top of the planting, which left the area looking as if a hurricane had passed over it. First thoughts of newcomers were always, "How can the banana plant force its way through that mass of debris?" which often covered the ground to a level of four feet.

In fact, the fallen vegetation provided protection for the seed; fire did little or no damage, nor did a drought lasting several weeks. The main factor was that the banana plant had no branches and the first sprout emerges from the ground as a tightly coiled leaf, perhaps an inch in diameter, and continues growing around logs and branches on the way to sunlight. As it reaches the top, the coil opens and the plant displays two leaves, after which the increase of growth and the additional leaves is rapid until at about nine months, when the fruit emerges from a crown of not less than eleven gigantic leaves.

Such was the "new land" planting, as we called it, and for which I went to Guaymas to help supervise. I had my first experience in Farm 23 which, being against the hills on one side of the district, included all plantable land, thirteen hundred acres, which was not enough for two farms and a bit too much for one.

Guaymas District covered a big area and carried four parallel railway lines, sixteen hundred yards apart. The total planted over the years was over fifteen thousand acres. Unfortunately, the soil was acid and gave little resistance to the spread of disease. District headquarters was a large and comfortable building — eight bedrooms, four baths, an office, dining room, kitchen, and servant's quarters, plus a tennis court and a movie once a week; in short, all amenities available at that time. I enjoyed my stay there which was not long. There were never less than eight boarders and, on weekends, the fellows came in from the farms to visit, play tennis, gossip, and play cards. The superintendent had been transferred from British Honduras where an earlier planting had died out; a nice, decent chap and well-meaning, but that was all.

On completion of the planting of Farm 23 in early December, 1925, I decided to take a vacation in Jamaica, to stay with Karl and Aline, who had been married only a year. I mention that vacation only because of an unusual series of incidents which preceded my arrival in Jamaica.

United Fruit imposed one inflexible rule in all plantations: no laborer could refuse to turn out for work, regardless of weather, if notice was given to begin harvesting the fruit. Provided that the railroad was in operation, heavy continuous rain was of no consequence whatsoever. It was with some disgust that, on the last day before my departure on vacation, a Sunday, I woke to the sound of heavy rain. The evening before we had received instructions by phone to harvest many thousands of bunches of bananas to load on a ship already tied up to the wharf in Tela. I would be a passenger on that ship for Havana, Cuba, for there was no direct sailing for Jamaica.

I had to get away early that morning to make a fast run by rail car to many farms, to see that all had started harvesting. There would be no refusals, but there would be reluctance in moving from dry quarters to a cold shower which would last all day. Even the mules were loathe to move in such weather. I would be using a poncho type of raincoat all morning and would not be able to get at my gun even if I needed it, so decided to leave it in my room.

At six in the morning I went to the messroom and the table was not set, nor was there any sign of the messboy, a Honduran. The cook, a Martinique Negro, was there so I asked, "Where is the messboy?"

"He's still sleeping, sah," was the reply.

"You go tell him *hoy es domingo pero no es dia de fiesta*" (Today is Sunday but not a holiday.)

The cook laughed and roused the messboy, and in due course I got my breakfast. I don't remember now, but I probably greeted the messboy with the same words that I spoke to the cook with.

As soon as possible I left on my rail car. It was still raining and did not let up that day. About noon I had completed my rounds and, finding no hold-up, returned to the headquarters for lunch. I was met by the superintendent in a very nervous state.

"The messboy killed the cook with your gun and then shot himself," he said.

He pointed, "There they are under the office," which was built about five feet above the ground.

I got the whole story from the others. After I left the cook continued a flow of laughing gibes at the messboy, repeating, "*Hoy es domingo, no es dia de fiesta,*" and not noticing that the boy was becoming increasingly sullen. The cook did it once too often and, when the boy went to tidy up my room, he saw the gun on the table, and this must have instigated the thought of retaliation. He went to the kitchen and killed the cook and, on the steps on the back porch, stuck the gun in his mouth and killed himself. Such tragedies were common incidents in Honduras. The local *comandante* came to view the bodies and take statements from witnesses. There was no other inquiry, and late that evening I boarded my ship in Tela.

Havana was then the "fun city" of the Caribbean. I enjoyed two days there with another employee of the Company, also on vacation. We saw all the sights, especially the night life! Returning to the hotel one night, after midnight, we found the lobby full of people and police, all standing around and gaping at the body of the night clerk on the floor. He had commited suicide. Of course, my thoughts immediately returned to the messboy in Honduras, only three days before.

Next afternoon I took the night train to Santiago, near the east end of the island, where I was to take another boat to Jamaica. It was a pleasant ride with a comfortable bunk and good food. Late the following afternoon the train was approaching Santiago when suddenly the brakes were thrown on and the train ground to a stop. After awhile, with others, I got out and walked to the head of the train. For no reason that any of us could see, on a straight track with no curve in sight in either direction, the locomotive had struck an ox-cart and killed the driver. The engineer undoubtedly sounded his whistle long before the impact and the ox-cart driver must have heard and seen the train. The oxen just barely got across and were not seriously injured, but the driver must have been paralyzed with fear as he had made no attempt to get out of the cart. In Cuba such incidents were handled with a little more authority than in Honduras, as no one could touch the corpse or move the locomotive until the local judge appeared and concluded his examination. It was two or three hours before we continued the journey, but I did not miss my boat.

In the twenty-four hour trip to Jamaica I seriously pondered the series of fatalities which seemed to be dogging my footsteps. Should I join my sister and brother-in-law? Although I am not superstitious I found some comfort in the thought that such things usually happen in threes, so I felt reasonably safe. And so it was, for I have nothing to report of my five-week stay with Karl and Aline.

I returned to Guaymas in early February, 1926. With the exception of an old and grouchy conductor, Mr. Hunt, who called me a liar and pulled a Derringer on me (he was moved to work in port that same day), there was a pleasant, friendly lot of fellows at Guaymas in those days. Lester Bragg, a nasal-voiced New Englander, was older than the average and we called him "Colonel," although I don't know why. Davis was another engineer I liked very much, but he had personal problems, I believe, as he never joined us in the "gossip" club after supper. Hardwick, a Canadian, soon left us under a cloud, and Henry Hogaboom, my good friend for many, many years, who worked on the new land plantings with Jack Roth. Jack was a bit of a character. For many years he handled all the new plantings in Honduras, which means that there are few individuals in the world, if any, who had brought more forested land into cultivation without any mechanical aid whatever. He was very good at handling large numbers of men, three or four hundred on piece-work or contract, given out to contractors with twenty to eighty men. Using only hand tools, they felled the forest, dug the ditches, and cleared the logs from the mule pack roads. Jack never rode a mule unless he planned to be away all day at a distant job; his supervision was done on foot. Thus, he missed nothing but spent a hell of a long time on the job, literally from dawn to dusk.

Of course, as the underbrushing of the forest proceeded, the animals and birds retreated before us. Occasionally, before sunrise, if there was a reason to be out at that hour, I saw a *danto* (tapir) ambling along his accustomed track to the nearest water deep enough to take refuge. There was a partridge-type bird to be found at times but very difficult to shoot among the trees, and wild turkeys, or *pavas*. But the only *pavas* I shot were too tough to eat.

Whenever Jack Roth started on a new farm, we would ask him to leave a sizeable area of grass untouched until the last. Such areas did occur but not often in the coastal rain forests. Jack would arrange to start underbrushing on all sides of the grassy area, approaching closer and closer, day by day, levelling all bush and trees of less than three inch diameter to facilitate the lining as described before. At least some of the animals in the area retreated to the grass patch every night. When only that area remained untouched, Jack would advise us to arrange the hunt for the next day. Our close friends were invited, but any laborer owning a shotgun or an old government rifle invariably showed up as well.

On the appointed morning, Jack would put in about fifteen men on each of the three sides of a five-acre patch of shoulder-high grass, and start cutting it low with machetes. We, the guns, lined up about forty yards away from the grass on the fourth side. There would be much shouting and racket as the chopping progressed and the men would catch an occasional armadillo too stupid to move away in time. The animals retreated away from the disturbance and, unknowingly, towards us.

One such such hunt the first animal to break through was a *tepescuintle,* which was promptly killed by the nearest gun. Then came three small forest deer, one after the other, and at least eight wild pigs, collared peccaries, burst out together to a fusillade of perhaps ten shotguns and rifles, everyone shooting at once. Why no one was ever killed, I don't know. I shot a pig which was claimed by another and I readily yielded it to him as, unless very young, collared peccary is tough and rank. I will agree that it was not a very sporting hunt, but it should be remembered that the only other way to hunt in a rain forest is at night with a flashlight, which is not sporting either. Dogs are useless in thick forest except in hunting *tepescuintles* which are inclined to take refuge in hollow logs if hard pressed and cannot get to the water.

In the foregoing section, I described the underbrushing and planting of banana farms and the occasional game hunt which occurred. Next came the felling of all trees, including some monster silk cotton trees of perhaps six feet in diameter, all axe work as there were no chain saws in those days. The felling created a spectacle which only a hurricane could duplicate, a mass of debris six feet high in some places and seemingly an impossible barrier to anything planted below it. Perhaps impossible to some plants, but not to bananas for the leaves and twigs soon dried and fell off, permitting some light to filter through. The first shoots from the planted bulbs emerged tightly coiled and did not open until well above the debris, after which the growth was rapid.

Bringing a new farm to production was not very exciting but entailed a tremendous amount of work, and it usually took ten to twelve months before the new bunches were ready to harvest. Shortage of labor was usually a problem since, of all the farm practices, only harvesting paid good money. Usually rates were paid enabling a cutter and a *mulero* to earn from two to four dollars each for a day's work, high wages which were only exceeded in the Caribbean by the Cuban cane cutters during six months of the year. But farms had to be kept clear of brush which thrived in the sunlight and frequent rains; the exuberant growth of the bananas produced many suckers or followers which had to be reduced to one for the next generation by frequent pruning. Roads, at specific intervals, were cleared of logs and intersecting ditches were bridged in readiness for the first harvesting. At this time a "cutter" selected a bunch of fruit at the right grade of maturity, his "backer" carried it on his shoulder to the next mule road, where the *mulero* loaded two mules with four bunches of bananas and took them to the railway, where they started their journey to the States by way of a UFC port and ship.

Perhaps I should provide rather more detail of the harvesting of bananas, the most important day in the life on a banana farm, and the most important function in any agricultural enterprise. Certainly in the early days, the cutter was the skilled man of most farm operations and is no less important today. Unlike the harvesting of most fruit in the States today, the banana must be harvested at a precise grade of "fullness," or maturity — mature enough to provide the heaviest weight on arrival at destination, but not mature enough to ripen on board ship in the five-day voyage to the States, or a thirteen-day voyage to Europe. So it was the judgment of the cutter and his experienced eye which determined the selection of one among many bunches in view. His judgment also provided the most accurate estimate of bunches available for harvesting in the following week in the area alloted to him — an estimate of great importance which is detailed elsewhere in this narrative.

The cutter carries a light, eight-foot pole armed at one end with a broad socketted knife. Of dozens of bunches visible, he selects one of the required maturity, probably a heavy, eighty-pound bunch hanging twelve feet or more above the ground. I am writing of the Gros Michel variety which was commercially popular fifty years ago. Having selected his bunch, the cutter reaches with his pole and jabs with the knife at the relatively soft fibers of the plant stalk, below the level of the hanging bunch whose weight has already inclined the entire plant slightly toward the ground. It takes but one or two jabs in a plant stalk, which has no woody fiber at all, and the crown of gigantic leaves with the bunch bends over and rapidly falls towards the ground. Then the backer steps in to receive the bunch on his padded shoulder and the cutter cuts it loose from the plant. The backer shoulders the fruit to the nearest farm road and leaves it to await transport to the railway. Meanwhile, the cutter fells the parent stalk close to the ground where it soon decays and provides fertilizer for its "follower," a selected shoot which may already be ten feet high or more.

Such was the harvesting of bananas in my young days.

There was no "fooling around" in the estimates of production for a specific day or week. An overseer or superintendent could be hauled away from a bar to phone in a production estimate for the next day or week and would be judged by the accuracy of these estimates, for they furnished a basis for the shipping allotted for each period. I doubt if computers have been able to take over this task yet.

On the newly planted farm the overseer was judged by his ability to get and hold labor, and keep his farm free from bush and properly pruned. Both practices are essential in the production and efficient transport of heavy bunches from the plant to the railway. Once in full production, his three-week estimate of the number of bunches his farm would produce in each of the weeks was of special importance. His estimate for the second week was a vital statistic in the aggregate of all such estimates over perhaps twenty-five thousand acres of bananas. This aggregate, after approval and/or adjustment by the Superintendent of Agriculture and the Manager, was radioed to the UFC headquarters in Boston, where it formed the basic allotment of shipping to the division for each week.

I found much interest in the estimates of production — weekly, triweekly, and monthly — for the budget and annual reports. Experienced overseers and superintendents, if interested, found their eyes continually roving about the bananas to note the hanging bunches in their daily rides about the farms. With experience, perhaps subconsciously, they learned to apprecite any marked increase or decrease of fruit nearing the required degree of maturity. A roving eye was permissable on horseback but later, in Tiquisate, when the superintendents were given cars and all the farms were equipped with gravelled roads, I remember Myrtle, by then my wife, saying that she hated to ride with me through the farms as my eyes were never on the road! Only the scarcity of traffic on those internal roads made that possible.

In the early days from 1926-30, my interest in estimates got me into trouble. As in all agriculture, the production of bananas was affected, sometimes seriously, by temperature and rainfall. In Honduras, in the district relatively close to the coasts, we could expect "northers" between December and February, periods of three to five days of strong winds, rain and low temperatures. A drop of temperature overnight from 75 to 55 degrees was not at all unusual with a norther, and always accompanied by cloudy weather. It seemed to me that such periods obviously affected the first appearance of the bunch of bananas when it "shot," was we called it, from the crown of leaves. If so, all such "shooting" in many thousands of acres would be retarded with a consequent effect three months later in the period of high production. Nothing would be lost, but certainly retarded and, as I reasoned, would cause a decrease in production if only for a week.

The district offices kept daily records of rainfall, and temperature and humidity were taken twice a day. To this I added a daily note of "full sunshine" or "marked cloudiness." After records had been carried for three months, I began comparing the production of each week, shown by graphs, with the weather records taken thirteen weeks before, the average time elapsed for a bunch of bananas to go from the shooting stage to a marketable grade of maturity. Sure enough, any serious drop in production coincided with a period of lower temperatures than usual and a lack of sunshine thirteen weeks earlier.

I was rather proud of my records and thought they could be useful in making monthly and weekly production estimates. I probably talked too much about it, which nearly caused my downfall. There came a period when production fell abruptly, and we failed by many thousands of bunches to load a ship. In fact, I believe it left half-empty for another division. My records showed that a decrease of production was probable, and I may have talked too much about that.

Early the morning after we had failed so dismally to meet our quota, Mr. Chute, Superintendent of Agriculture and my boss, came to the district of which I was assistant superintendent.

"I want to ride over your farms," he said. "I can't believe you got out all the available fruit."

This was a serious crime as mature bananas lasted very few days on the plant before ripening, and this was much too late for export. I was not particularly worried because the weather had been fine and there was no reason for the fruit to have been left. My frequent inspections showed that overseers and foremen were competent and watchful.

We rode on horseback through the bananas for five hours, mostly between the roads where lazy cutters could have left a few bunches. We found practically no available fruit.

Then Mr. Chute spoke his mind. "Hutchings, I came out here to fire you because I thought your talk had convinced the overseers of a drop in production, that they had passed it on to the foremen and cutters, and nobody really looked for the available fruit. But I can't find any in the farms, so we'll say no more about it."

I was shocked as I had not expected to hear that. I guess I did not mention my records and estimates any more, but I still believe they had something of value.

Another Promotion

In the period of 1925-26, I was transferred from assistant superintendent of the Guaymas District to acting superintentent of the Ulua District. I do not recall the reason for the transfer or what happened to the previous superintendent, but the transfer indicated appreciation of my work in Guaymas under a superintendent who was rather a nonentity, although pleasant and easy to work with. For this appreciation I was indebted to Earl Ames, the recently appointed Superintendent of Agriculture, on transfer and promotion from Guatemala. He was a tall New Englander, whose light blue eyes could be truly frosty on occasion.

Mr. Ames was the first man I had met from the higher ranks of the Company, whose thoughts were concerned with the opportunity and necessity for higher production per acre, especially in short-life land. Our senior superintendent, Ed Cobb, was a driver rather than thinker, and all his training had been under a Superintendent of Agriculture who rarely left his office in port and whose only talent had been rigid control of maintenance expenditures, and disinterested if it cost money. I well remember Mr. Ames' first query for information, directing our thoughts toward his quest for higher production. It was, "What happens to the fruit?"

A very pertinent query for we maintained a mature population of banana plants of five hundred and fifty per acre, and were well satisfied to ship two hundred and twenty-five bunches per acre. The heaviest losses, from high winds and Panama disease, were uncontrollable, but, from that query, in time, we found an optimum population of four hundred and fifty plants per acre for our lands and rainfall, which resulted in a production of three hundred and fifty bunches per acre on several farms.

So, I became a great admirer of Mr. Ames, and cannot emphasize enough the importance of his training and influence on my thoughts as I gained experience. In later years I visited him in his retirement in Homestead, Florida, as often as I could and felt very proud indeed when he told me he had considered me "one of his bright young men."

The Ulua District was comprised of seven farms, all bordering on the banks of the Ulua River, and totalling more than five thousand acres in bananas. Naturally drained, the soils were sandy and clay loams with a high pH, this making them tremendously productive in banana culture. On Los Indios Farm, the harvesting of sixteen-hand bunches, weighing 120 pounds or more, was commonplace. The difficulty was not in growing big bunches but in harvesting them and lowering such heavy bunches to the ground without breaking them. This was solved in time by one of our inventive overseers.

Of the seven overseers, four were Latin. Don Pancho Belmar, a fair, good-looking Mexican, was on Buena Vista Farm, and one of the laziest overseers I ever knew. Too lazy to get off his horse to urinate, he would merely open his fly and let go, wetting his pants in the process. Don Marcos Galvez was a Nicaraguan, and husband to one of the most obese women I ever saw. Not more than five feet tall, she weighed more than two hundred and fifty pounds. A special chair had to be made for her as she had broken too many of Company issue. I recall nothing of the Honduran overseer on Palo Blanco. The overseers of Los Indios, Uraco, and Palomas, were American, one of whom, Fred Warren, became my good friend. Of the overseer of La Fragua Farm, I shall have too much to say as he was the villain of the incident I shall now relate.

A Close Call

Don Gustavo Pinel, Honduran, was a full-blooded Indian, or nearly so, evidenced by his black eyes and lank, black hair, his coppery complexion, and very sparse facial hair. His antecedents were cloudy. He hinted that he had been a confidential Company employee, engaged in the acquisition of banana lands, sometimes by rather nefarious means. Before my time, he had been employed for long periods during which he reported directly to the manager. He hinted that he was allowed bodyguards because of services performed for the Company. Pinel was a man who may have held that post as he was always around and did no work. I did not dislike Don Gustavo, although I did not care to lunch with him and his family, as I felt his invitations were more for the purpose of building up his own ego than providing hospitality. His farm was satisfactorily handled and, if it was due more to the efforts of his foreman than his, I had no quarrel with that. As I remember it, he had no personal friends in the district. In my presence, he was positively genial when talking to his men, but I felt that was not normal.

In those early years, Boston, head office of the United Fruit, interfered very little with the agricultural practices of the six divisions which were the principal producers of bananas. The head office watched the costs, of course, but production was variable in countries subjected to floods in the rainy season, and high winds at any time. Such disasters were expected and, indeed, the budgeted production was always much higher than the sales estimates. In a very prosperous enterprise there appeared to be no reason to criticize or suggest changes in the agricultural practices of these successful divisions.

Therefore, we were very surprised to receive instructions from Boston to establish a sizeable experiment on all farms — twenty acres, I believe — in pruning bananas. Whose brain wave it was I do not recall, if I ever knew, but the instructions were ridiculous and reduced production due to the time-limiting experiment, and caused the loss of many thousands of bunches in the Ulua district of seven farms. I protested immediately to Mr. Ames, who replied, "I know, I fully agree with you, but we were not asked for opinions, just instructed to establish the experiments."

On every farm, I carefully explained the experiment in the field, and pruned a small sample area, as instructed, for the experiment. The overseers were unanimous in condemning the value of the experiment, but I could only reply with the words of Mr. Ames.

After several months in the district, the overseers were fully aware that I made it my business to supervise everything of importance or interest, which affected the production. My first inspections of the experiments on several farms passed without comment. The instructions for pruning were simple and there was no room for mistakes, so I was astounded when I got to La Fragua and saw Don Gustavo's experiment. It was being "butchered" and I was immediately convinced that it was deliberate and that Don Gustavo had decided to demonstrate his contempt for instructions. He had behaved in a similar manner with an earlier acting superintendent, to whom he said, "You can stuff your instructions," and got away with it too!

To some extent I "blew my top" and made the fundamental

mistake of calling to the men to stop working. I should have had Don Gustavo do that; I had caused him to lose face in their presence. I was so mad at what I felt was deliberate provocation that I dared not speak to him for several minutes, and then questioned him but got no replies except for his opinion of the stupidity of the experiment. I did not have lunch with him that day, nor did he invite me.

After consulting with Ed Cobb, the senior superintendent, I went to port the next day for an interview with the assistant manager and explained the situation to him, saying that Don Gustavo had showed his contempt to a previous superintendent, who had swallowed it, and now had tried the same thing on me and I was not going to take it. I felt that the existing situation could easily deteriorate to something quite serious if he remained an overseer under my supervision; but here I overestimated him as he was too much of a coward to face up to anyone directly.

Mr. Thomas, the acting manager, was sympathetic but showed that some substance existed behind the rumors that Don Gustavo enjoyed special privileges.

He said, "You know Don Gustavo is Mr. Goodell's man and he's away at the moment, but I shall take it up with him at his return, to relieve you of Don Gustavo's vagaries. Meanwhile, go easy with him."

Perhaps two or three weeks passed when I received instructions to clean the main drainage ditch which ran parallel to the river and drained the entire district. An important part of this ditch, the outlet, bordered La Fragua and I sent written instructions to Don Gustavo to put at least twenty-five men to work on it. I had not ignored him in the meantime, nor had I accused him of anything. I was content to wait for the manager's return so it was a kind of truce, as he suddenly became unusually attentive, as if he sensed that he had gone too far and wished to gain approval.

After some days, I phoned Don Gustavo and told him I wanted to see the work in the main ditch and would come next day at about one p.m. I arrived on time and, on meeting him, suggested that we go in my rail car to just beyond the limits of the farm and walk from there to the ditch, more than a mile from the railroad.

But I have two mules saddled for us," he replied. "It is better that we go on mules."

"No," I said, "we can only ride to the same place by either

mule or car and from there we must walk, as you know."

He was insistent, "But I have something you should see on the way," naming some quite unimportant work. He continued until he realized he could not change my intention. I wanted to spend only an hour or so inspecting the ditch and had arranged to meet the overseer of the next farm that same afternoon. So we went by rail car to a subsidiary ditch which drained into the one I wanted to see. I believe it was fortunate I did not change my plans to make a quick inspection, as I consequently frustrated the plan Don Gustavo had prepared for me.

Cipriano, my motor boy, took orders by phone to run to Kilometer 54, where there was a railroad spur, and a telephone to receive his report of arrival. There, Don Gustavo and I left the railroad to walk to the ditch. We walked along the dragline right of way, cleared when the ditch was dug not long before. I followed the path made by the men who worked on the ditch every day. For reasons of his own, Don Gustavo started to lag behind and take the more difficult route at the edge of the clearing, perhaps fifty feet from me. He also fired his revolver several times, at nothing I could see.

As we neared the ditch, he lagged still farther behind and I stopped to wait for him. I suppose I had become subconsciously uneasy as I was alone with a man of dubious reputation and who had, I believed, tried to challenge my authority in the district. My thoughts perhaps flicked back to his insistence that we ride mules instead of the rail car, which was quite out of pattern. The walk from the railway had been marked by his silence and his loitering behind and choosing a more difficult route at some distance from me, all of which was certainly not threatening, but strange in the circumstances. I have been told, more than once, that I am unusually sensitive; if so, I believe it served me well that day.

We neared the ditch which was less than a hundred yards away. Two men were sitting there and got up and started walking quickly towards me. There were no others working and Don Gustavo had definitely stated that more than twenty men were working there. Then, in a flash of intuition, I knew! Don Gustavo had set me up to be killed by those men.

I turned about immediately and started walking back as fast as the terrain permitted, and I do mean fast! As I neared Don Gustavo, I deliberately drew my gun and half-turned to face him, about twenty yards to my right. He did not look at me and his pistol was holstered. I said nothing, nor did he open his mouth. An innocent man would surely have made some surprised comment at my precipitate return and non-inspection of the ditch, with gun drawn. Was I unreasonably fearful? I cannot accept that now.

I believe his murderous plan went wrong because I insisted on taking the rail car to K. 64 instead of going on muleback, for Cipriano was then a witness that I had left the railway with Don Gustavo. If we had ridden mules he would have given some plausible statement that an urgent message had followed him from his wife, and that I had impatiently decided to go on alone. Quite likely in my stubborn capacity, and Don Gustavo would have readily produced the witnesses to prove his recall.

I walked all the way out until it was clear that the men were making no effort to overtake me. The danger was over, for a man with a gun has the edge over two men armed only with machetes, unless taken completely by surprise, as was intended. On reaching the railroad I moved off as quickly as Cipriano could get orders from traffic control, leaving Don Gustavo to walk the two miles to his house, which I am sure did not sweeten his thoughts. I never did see the ditch, but arranged for the inspector of farms to do so when Don Gustavo's early transfer required a routine payroll check of work completed and in progress.

Did I let my imagination run away with my common sense? Perhaps. Certainly I did not have a shred of factual evidence of intent. Time smooths all vivid impressions but today, more than fifty years later, I still count it as one of the incidents in which only good fortune saved my life.

On return of the General Manager, Mr. R. H. Goodell, Don Gustavo was moved almost immediately to an obscure job concerning Company land titles and boundaries. He was given a rail car which he doubtless considered a status symbol and soothed his ego. I told only two people of the incident, Ed Cobb, the senior superintendent, and my good friend, Octavio Schofield, the latter having vast experience in handling men in Latin America. Both agreed that a plan had been made to kill me, which meant their assessment of Don Gustavo's character and capabilities concurred with mine. However, Tavo reminded me, "You were not blameless, you know. You caused Don Gustavo to lose face in the presence of his men — something which he found unbearable."

Years later I passed Don Gustavo in the street of San Jose, Costa Rica. He looked at me with eyes as friendly as a snake's but said nothing. Neither did I greet him. I heard later that, when he was dismissed from the Company, he tried to blackmail them by threatening to publish his memoirs. When told to go ahead and publish, no doubt he realized that, if he had done any dirty work for the Company, he himself was not immune from the consequences, so he did not have the guts to publish anything. Thus ended my episode with Don Gustavo.

A Letter from Puerto Castilla

Truxillo Railroad Company
Puerto Castilla, Honduras
December 21, 1923

Dear Folks-at-home,

We have been in Puerto Castilla just one week today. I know it is not customary for one to begin to tell about their trip dating backward, but then, women have the reputation of being inconsistent, so why not get in line?

Not only did we arrive on Thursday last but we had to settle into our new home right away. My husband changed his bachelor quarters for a larger domicile and it has taken me all of the week to get adjusted, acclimated, and in fair running order. Unfortunately, I am one of the kind to whom it is very upsetting if the household is askew. But our house is ideal for this part of the country, being concrete throughout — floors, walls, etc. Fact is, one may take out the furniture if they wish, attach the hose, and the cleaning is all done. I know stone floors make you shiver but I haven't the least compunction in the world to paddling across them barefooted. So you may gauge the temperature here by that remark.

The house has just been furnished with Honduras mahogany and has all of the things that are available at the Company commissary. This means that we are short this, that, and the other, particularly dishes. But we are a family of New Englanders and "make-shifting" comes natural. The baking powder biscuits were cooked in the frying pan and the cake in an enamel bowl. Curtains are not customary — they shut out the air — but I was not content until I had made at least half-curtains out of gauze.

The porch of our house is also the living room. All openings are minutely screened, but even the finest mesh cannot keep out a creature they call a sand fly. I believe he is even a greater pest than the mosquito. It is infinitesimal, but oh my! We are absolutely obliged to sleep in beds canopied with a net or we should be eaten alive. It seems like sleeping in a tent and sometimes I wonder where I am when I wake up.

Puerto Castilla is a new concession for the United Fruit Company. I would say it has made remarkable progress when comparing the present layout with the pictures I have seen of the earlier stages. It is all man-made land consisting of sand pumped in from the shore and is virtually a small island. Behind us is a canal from which may emerge crocodiles and other foreign reptiles. Along the other bank, blue herons and long-legged white storks sit and fish. The ugly black buzzard is ever hovering overhead to see whom he can beat to get a meal. The big fish chase and "gollop" down the baby shiners. Never have I seen a better illustration of "might makes right" than goes on behind my back door.

Within a few minutes walk is the ocean front and truly the bathing is delightful. Back in New England, one has the habit of dangling along the beach and going in by the inch, at least I do, but here one may plunge right in without a shiver of cold. The water is particularly buoyant. Sunday last, I saw two little shavers down on the beach. They were the Riggs boys, ages three and five, each attached to a pair of water wings bobbing up and down on the waves like bits of cork. They cared little whether their feet touched the bottom or not. Perhaps even I shall learn to swim.

Across the street is the park which will eventually be a sort of botanical garden. It has been grassed down and has a fountain in the center. Each pathway is bordered with hibiscus, both double and single, which has a beautiful deep pink blossom resembling an enormous fuchsia. They irrigate this choice spot to keep it in condition but I don't know why. Every day since I have been here, the heavens suddenly open up and spill down buckets of rain. Never have I seen it rain harder or more unexpectedly. One minute the sun is shining and the next we are deluged. Fortunately, this is supposed to be the end of the rainy season. When I exclaimed to someone who had lived here the past nine months, "Aren't these showers beastly?," she said, "No, I don't think so. It has been so dry and hot and dirty that I like nothing better than to stay right out in the rain and let it pelt down on me." But a different viewpoint is what makes life interesting, isn't it?

Getting meat, vegetables and groceries is just as difficult as obtaining dishes and furniture. Fresh vegetables are a gamble. They will send you a box from up the line but this box may contain anything — not what you want or ask for. You pay for it just the same and try to use what is there. I have not yet tackled the native vegetables. We did have a strange bit of fruit, called papaya, given us for dinner the other day. It belongs probably to the melon family but tastes like pepsin gum. I think that I shall get to like it as I am partial to all kinds of fruit.

A New Orleans boat is due in once a week bringing butter, eggs, and some fresh vegetables. Otherwise, you buy in tins from the commissary. I have learned that it is well to hang around as soon as you hear that the goods have been released from the customs, for it is first come, first served. I have been waiting eternally for creamery butter and go every day to ask about it. Usually, I am placated with a little tin of Montequillar which is butter *a la* something else.

About everything is subject to duty as the Hondurans depend on taxation of this kind for their revenue. I paid 75 cents for a bottle of vanilla flavoring and 40 cents for a little tin of raisins. I am about to go after confectionary sugar so that we may have frosting on our cake. This too will be in the luxury class as is Domino sugar that comes from the States. Granulated sugar, a coarse yellow product, is only 10 cents so I am not getting my money's worth even in that. Bananas, oranges, grapefruit, and pineapples can be had very reasonable, but it is the same with them — feast or famine. I have seen the bananas being loaded onto ships, but none to buy.

Beef is native and, like the Honduran chickens, runs its legs off down to the bone and muscle. One has to stew up most things and make them up into delectables.

Now, you will know what I have been doing with all my time. In between experiments, I am teaching school. I'm not sure that the teachers in Shaw School will approve of my methods but I'm polishing up. Meanwhile Katherine and Newcomb think it is one grand lark down here.

(Writer unknown)

This letter was passed from Harriet and Eddie Couture to Millicent and Scotty Crichton. In May, 1983, the Crichtons made a copy for Clyde Stephens in Fort Myers, Florida.

Puerto Castilla, Honduras, was the headquarters and port for the Truxillo Railroad Company. This United Fruit Company town was built on sand and mud dredged from the swamp (left) and bay. Note the huge hospital in the foreground. The entire division was abandoned in the late 1930's and literally grew back into jungle again. (Photo: Unifruitco, April, 1931)

The mud and sand landfill at Puerto Castilla, Honduras, was transformed into a community of gardens and parks. Ladies in the park are: (top row) Mrs. Miriam Walker, Miss Phyllis Arscott; (middle row) Miss Gertrude Smith, Mrs. George (Vivaneen) Rippey, Mrs. Clarence Holdsworth; (bottom row) Miss Rosalie Bayne, Miss Ruth Schwartz, Miss Lillian Purrington. (Photo: Unifruitco, September, 1929)

Detour to Tela

By

Dave Crichton

In 1929, I was temporarily assigned to work in San Pedro Sula, Honduras, at the Ulua Brewery and Ice Plant under the management of Colonel Guy Molony, better known as "Machine Gun" Molony. He was the *comrade-in-arms* of the famous soldier of fortune in Honduras, General Lee Christmas. The arrangement for my services was made by my employer, the Tela Railroad Company, a subsidiary of United Fruit Company. I had been in San Pedro Sula for about three months when we noticed that our supply of ammonia gas was running short. Colonel Molony contacted the Tela Railroad Company headquarters in Tela and arranged for four cylinders of gas to be released to the brewer of *Cerveza Lempira.* He also made arrangements for me to go to Tela to pick up the gas. As the same time, I could have a visit with my family living there.

A trip to Tela was not usually a quick one because the rival Cuyamel Fruit Company on the west side of the Rio Ulua had not yet been connected by a bridge crossing to the Tela Railroad Company on the east bank. Also, at this particular time, there was considerable unrest in the country due to the Hererra revolution and other disturbances.

I left San Pedro Sula on the *pasajero* — passenger train — for Tela and after three hours, arrived at Baracoa on the bank of the Chamelecon River. Here, you usually transferred to a motor launch which went down river to the Chambers Canal, then into the Ulua River, continuing on to Melcher Junction where you transferred to a train from Progreso bound for Tela. This time, the usual connections could not be made. There was a contingent of government troops on the river bank and we Tela-bound passengers were herded back onto the train and taken to Puerto Cortes, the banana port of the Cuyamel Fruit Company of Sam Zemurray.

David Crichton was born in Edinburgh, Scotland, in 1905. His father, David, Sr., came to Tela, Honduras, in 1919 after serving in World War I, leaving his wife and four children behind in Edinburgh until they joined him in 1925. Like his father, young Dave worked in refrigeration in the machine shop and power plant. In Puerto Armuelles, Panama, he worked in accounting and in the hospital. His career ended in 1964 in New Orleans where he worked with the Medical Department under Dr. Lawrence Drennan. Dave lives in Fort Myers, Florida.

Upon reaching the end of the line, things were not normal. Puerto Cortes was full of troops and confusion reigned. I couldn't find lodging or a room in Company quarters, but eventually I did manage to get a room in the Cozensas Hotel which I had to share with Jack Card, a railroad employee of the UFC in Tela. I stayed there for three days and two nights. A locomotive, with some four or five banana fruit cars and about a hundred or more soldiers, shooting at will from the roofs of the rail cars, shunted up and down the railroad tracks right under the balcony of the hotel all night long. It was a relief when Colonel Molony got word to me to get on the train Wednesday in Puerto Cortes and disembark again at the railroad station in Baracoa where I would be met. A motor launch would take me to Melcher where I would stay the night, ride a mule from there to Toloa Junction and be there in time to catch the train from Progreso to Tela. Fortunately, I made all of these connections despite the complications.

Upon my arrival in Tela, I found that all the women and children had been taken aboard the S.S. *San Gil* for safety and taken out to sea. The men were guarding the houses and property and standing by in case of fires. On Friday, I was assigned to guard duty, but by Saturday, things began to clear up. The women and children came back to port and folks returned to their normal activities again.

As planned, I managed to get the four cylinders of ammonia gas back to San Pedro Sula without incident and Colonel Molony was able to continue producing ice and Lempira beer again. Such items were necessities of life in those days.

Tela Railroad Company headquarters at Tela, Honduras. The horse race track circled part of the airfield (top left); the golf course was behind the Company town (upper left); old Tela (lower right). United Fruit started the banana port in 1914. (Photo: UFC research files in La Lima)

Tom Wolfolk (white suit) boarding the Tela bus. Such rail vehicles transported school children from remote banana farms to the American school in Tela, Honduras. Overseers and their families also came in to port for medical care, shopping and recreation on these motor cars. (Courtesy of Scotty and Millicent Crichton)

Parents and children of the Tela American School, Honduras, about 1934. They are: 1. Ma Grobe, 2. Buddy Grobe, 3. Donald Ames, 4. Sylvia Keen, 5. Blanton Evans, 6. Bufa Roberts, 7. Mrs. Doubleday, 8. George Bryant, 9. David Crichton, Sr., 10. Hazel Roberts, 11. Mrs. George Evans, 12. Bob Roberts, 13. Mrs. Thompson, 14. Mrs. Tom Wolfolk, 15. Sid Keen, 16. McMullen, 17. Ann Crichton, 18. Camille Doubleday, 19. Billy Grobe, 20. Rev. Auler, 21. Zona Grobe, 22. Margaret Thompson, 23. Billy Aycock, 24. Willene Grobe, 25. unidentified, 26. Nan Crichton, 27. Tom Wolfolk, 28. Woody Thompson, 29. Frank Doubleday, 30. John Doubleday, 31. Billy Wolfolk, 32. Myrtle Bryant, 33. Jean Crichton, 34. Jean Beasley, 35. Hinky Bryant, 36. Arturo Doubleday, 37. Louis Doubleday, 38. Nelda Beasley, 39. Ian Crichton, 40. Doubleday, 41. Alfredo Keen, 42. Bobby Bryant. (Courtesy of Nan Crichton Dahlgard)

Aviation caught on quickly in remote banana zones in the 1920's. Charles Lindbergh and Pan American Airways promoted flights throughout the Americas, including a stop in Tela, Honduras. UFC flew passengers and cargo in its Company Fokker, the ***Tela,*** *between Tela and Tegucigalpa in 1928, reducing the rugged 48-hour trip by land to two hours by air. This photo shows Lindbergh in Toncontin airfield in Tegucigalpa, January 5, 1928, during his "Lone Eagle" —* ***El Aguila Solitartia*** *— solo flight from the USA through the Americas. (Courtesy of Virgilio Moncada)*

High society at the manager's party in Almirante, Panana, about 1920. (Collection of Clyde Stephens)

Memoirs of a Naturalist

By

Alexander F. Skutch

Before World War II, scientists who wished to study in the Caribbean countries would have found it much more difficult without the United Fruit Company. They traveled on ships of the Company's "Great White Fleet," often with complimentary tickets, found hospitality at the Company's research stations and farms, and, if they fell sick, sought aid in the Company's hospitals.

The botany department at John Hopkins University, where I did my graduate work, made periodical pilgrimages to Jamaica, then a British colony, to introduce the students to tropical vegetation. In those days, graduate students were often short of money, and three of us did not have enough for the Jamaican trip in 1926. Dr. John R. Johnston, the Company's director of tropical research, came to our aid, offering free transportation to and from Jamaica in return for making certain studies of the banana plant. I chose the anatomy of the leaf for my research, and, after the rest of the party returned to the States, I spent nearly two months on Platfield Plantation in St. Mary's Parish. There, I set up a field laboratory in a timekeeper's cottage, with a cook and houseboy to take care of me. My cook nearly ended my career prematurely by serving me a dish of akee that had been improperly prepared, for which she received a severe scolding from Mrs. Keith, the tall overseer's little wife. Despite a brief illness, at the summer's end I returned to Baltimore with material for my doctoral dissertation.

Two years later, when I had received my degree and a research fellowship from my university, Dr. Johnston arranged

---

Alexander F. Skutch was born in 1904 in Baltimore, Maryland, and was destined to be a naturalist. He first came to the Tropics in 1926 on a United Fruit Company ship as a student on a botanical expedition in Jamaica. After getting his Ph.D. in botany from Johns Hopkins University, he worked for the United Fruit Company in banana research in Panama. Dr. Skutch studied the fauna and flora of Central America and has written numerous articles and books. He has received many prestigious awards and fellowships for his contributions to understanding nature, especially as an ornithologist. Dr. Skutch married Pamela, the daughter of C. H. Lankester, another renowned naturalist of Costa Rica. The couple still lives in a humble home along the Rio Peña Blanca in the Valle del General in Costa Rica where they preserve the last one hundred hectares of rain forest in the region.

for me to reside, as a visiting scientist, with Joseph Permar, a veteran of World War I who was in charge of the one-man research station at Changuinola's Farm Six in the Almirante Division in Panama. Permar used to say, "We don't make much money, but we have lots of fun." He lived in grand style, in a spacious, well-appointed house in a beautiful garden, with a cook, yard boy, secretary, and motor boy to drive him along the rails in his little two-cylinder, open, red motorcar. Once a week the supply train from Almirante ran into the Farm Six spur and deposited his provisions at the front door.

Permar's assignment was to produce a commercially acceptable banana resistant to Panama disease then ravaging the plantations. He had a large collection of species and varieties of bananas gathered from all over the tropical world. By fertilizing the normally seedless Gros Michel variety with other kinds of pollen, he managed to obtain a few viable seeds. When, after months of waiting, the hybrids from these seeds fruited, the results were always disappointing. I do not recall whether any of his productions were disease-resistant, but that didn't matter because the fruit was so inferior. Nevertheless, history was being made because this was a pioneer effort in banana breeding.

For a young naturalist on his first visit to the continental tropics, my six months at Farm Six was a wonderful experience. The garden was full of birds, all new to me, some of which I studied while they nested. The research house faced the Changuinola Lagoon, an old channel of the Rio Changuiola, which wound for many miles down to the Caribbean Sea. On Sundays, Permar and I sometimes voyaged down the lagoon in a *cayuco,* paddled by Iriah Morgan, the yard boy, who had much nature lore. Waterfowl abounded along the lagoon, white-faced monkeys, howler monkeys, and three-toed sloths foraged in the trees along the shores, and silvery tarpons broke the surface. Sometimes we explored the great swamps of gigantic silico palms, *Raphia taedigera,* near the coast, a most impressive vegetative formation. On other Sundays we visited different parts of the region in the little red motorcar.

Sometimes I accompanied Permar to Flat Rock, a plant introduction station that the Company had on Columbus Island, where from the shore we could look across the water to Swan Cay, where boobies and other tropical birds nested. Thanks to the kindness of the Company officials at Bocas del Toro, I spent a memorable morning among these sea birds.

The following year, 1930, I wished to return to the Changuinola Research Station, but it was about to be closed, so I went to the Lancetilla Research Station, a few miles inland from the United Fruit Company port of Tela in northern Honduras. Here, Dr. Wilson Popenoe, authority on tropical fruits, was in charge, and Alfred Butler, a horticulturist from England, was his assistant.

Around the garden in the Lancetilla Valley rose steep foothills covered with unspoiled rain forest, where I became familiar with forest birds and plants that were absent from the cleared lands around Farm Six in Panama. One day I climbed far up into the mountains with Raymond Stadelman, who was in charge of the Lancetilla snake pit, which collected venom which was sent to Philadelphia for the preparation of antivenin. As we descended laboriously along a boulder-strewn torrent, we found a small *barba amarilla* or *fer-de-lance,* which my companion caught and placed in a white flour sack. Farther down, Stadelman captured, after a tussle, a large snake of the same kind. He insisted that I hold the sack, already containing one venomous serpent, while he inserted the other, tail-first. On the first attempt, the big snake jumped out, past my hands, to be replaced, also past my hands, on the second attempt. To travel with snake collectors was perilous.

After my six months at Lancetilla, years passed before I was again associated with the Company, this time as an employee rather than a visiting scientist. Through the kind offices of Mrs. Doris Stone, president Samuel Zemurray's daughter, I was chosen to superintend the newly established botanical gardens in southwestern Costa Rica at railroad Kilometer 53 at Esquinas and Piedras Blancas between Golfito and Palmar. They gave me a newly finished house on an elevation amid the forest overlooking the railroad and the gardens just beyond it. It was exciting to live amid so much wild nature; but the government restrictions made things more difficult than they had been in the old, unregulated days at Almirante two decades earlier. Getting provisions at the garden was particularly troublesome. I requested the commissary in Golfito to give me credit, but to open charge accounts even for Company employees was prohibited. Accordingly, I had to pay for my supplies when the morning train stopped briefly at the garden. One day, when the train started to move off before the man in the baggage car could give me my change, I told him to throw it. From among the ballast rocks, I retrieved most of the coins.

At the end of 1947, after nearly four months at Esquinas, I

decided that I would have to give up either the Esquinas Botanical Gardens or my farm at Quisarra in the Valley of El General, as traveling back and forth was very difficult then. I resigned my position and have not regretted it, for I still have the farm, but the botanical garden was abandoned long ago. My successor, Dr. Paul Allen, a taxonomic botanist, collected extensively and wrote a famous book, *The Rain Forests of Golfo Dulce.*

The old United Fruit Company has been the target of much criticism, but I am grateful to it and its research staff for helping me to start my lifelong study of tropical nature.

***Editor's note:** In 1959, the Pan American Highway survey gang routed the future road between the abandoned Esquinas Gardens and the forty-hectare forest where the huge overseer house was that Doctors Skutch and Allen had lived in. Just before this rare rain forest was about to be expropriated and destroyed in 1975, the editor bought the property as a forest reserve to honor the memory of these two great tropical naturalists. Today, located at road marker Kilometer 287, it is the only block of untouched forest left along the Pan American Highway in Costa Rica.*

Raising Cane and Growing Bananas

By

Mark Trafton, Jr.

As a Musician

My younger years were spent in Hanover, Massachusetts, south of Boston. In high school, I played the saxaphone and made good money playing in local bands. So much, in fact, that I cancelled several college applications while I was still attending Thayer Academy. After graduation from Abington High School in 1923, I joined the Elsmore-Nash Orchestra in Brockton and played almost every night at private parties in the public dance halls there. I acquired a clarinet and studied at the Boston Conservatory of Music to quickly learn to play the instrument. By 1924, I passed a difficult audition with the Paul Whiteman Arcadia Ballroom Orchestra in Providence, Rhode Island, and played with them for two years. Besides performing at the Arcadia Ballroom, we played a theatrical stage circuit, traveling to many popular theaters, which included two weeks at the grand opening of the Metropolitan in Boston. Early in 1926, I joined the Pearly Breed Orchestra at the Shepard Store Colonial Tearoom on Boylston Street in Boston where high society gathered. We also broadcast daily from the tearoom on Radio WNAC.

While still working as a musician, I enrolled in the Burdett Business College and studied typing and Gregg shorthand. My parents wanted me to give up being a professional musician and work my way up in business. So I think Mom and Dad "pulled some strings" to get my career going. A distant relative, Harold Cutler, had married Bessie Preston, daughter of Andrew Preston, co-founder of United Fruit Company in Boston. Also, my uncle Elisha Cobb and Andrew Preston had been life-long friends, so there were some "strings" to pull on.

Mark Trafton, Jr., was born in Pembroke, Massachusetts, in 1906. As a young musician, he played in the Paul Whiteman Orchestra and in the theater circuits. In 1926, Mark went to Banes, Cuba, and later became an overseer in the sugar cane fields of the United Fruit Company. He was a pioneer in the Puerto Armuelles, Panama, division where he landed in 1928. In 1942, the Trafton family was transferred to Honduras. Mark has a keen sense of the natural environment around him and pursued numerous interests during his 37 years in the Tropics. Besides his expertise as a banana man, he enjoyed tropical botany, birds, insects, fossils, photography and his amateur radio. After living in Boston, New Orleans and Miami, Mark and Harriet, his wife since 1935, retired in 1971. They now live in Tallahassee, Florida.

In September, 1926, I received a letter from the United Fruit Company asking me to report to Boston for an interview. They hired me as a stenographer in the Merchandise Department in Banes, Cuba, at the relatively high salary of $125 a month. A room would be furnished at no charge but I would have to pay $30 a month for meals in the Company mess hall. I was given a steamship ticket from New York to Havana for November 13th and a letter to the Company agent in Havana who would meet me there.

My Dad acompanied me to New York and we went to a vaudeville show at the famous Palace Theater where I had played an engagement the year before. That night, I had a tearful phone conversation with a girl friend, Mary Graham, in Providence. We had spent some time together when I was with the Whiteman orchestra but there was nothing very serious between us. But now that I was leaving the country, she was upset (and so was I for a few hours).

The next morning, I boarded an old Great White Fleet banana boat, the original *San Jose,* destined for Havana. The four-day voyage was pleasant except for the last night off Cuba when a severe tropical storm hit. It was rough and I was seasick and wondered if the ship would stay afloat. It did suffer some damage but we made it to the United Fruit wharf in the Havana harbor. The Company agent met me and put me on a government train from Havana to Entronque Dumois, the junction with the Company railroad near the northeastern end of the island in Oriente Province.

The two-day, one-night trip was interesting but the Pullman sleeper, a comfortble but old wooden car, squeaked and groaned all the way as it rocked along over a rather rough roadbed through endless miles of sugar cane plantations and pasture land as the steam engine spewed out clouds of black smoke.

At Dumois, I changed to the *Central Macabi* Company train for the short ride to Banes where I was assigned a room in the "bull pen" bachelors quarters. I was greeted by a few of the guys, most all English or Jamaican colonials and a few Cubans, and we easily became acquainted over a few *Cuba libres.*

Thus began a tropical venture which would be my life for the next 37 years before returning to live in the United States again.

Life in a Cuban Cane Division

The transition from my lifestyle in the States to that in Cuba was dramatic. In Banes, I found myself in a beautiful, warm tropical setting, with trees, palms and flowering plants completely new to me. Life there was in an orderly, quiet settlement typical of British colonial style. At the same time, it was in a Spanish-speaking country where the language and customs, outside of the Company compound, were completely different from what I had grown up with.

Banes was the headquarters of one of two large United Fruit Company sugar plantations in the Oriente (east) Province of Cuba. The other one, adjoining Banes, was Preston, named after the co-founder of the Company. The Banes Division included some 50,000 acres of sugar cane and pastures divided into about 35 farms in four districts. The large *central* — a sugar mill — was located nearby in another Company town at the port of Macabi where Company "M" boats (*Macabi, Mayari, Maravi, Manati,* etc.) loaded bulk raw sugar for refining at Revere in Boston.

The Banes Company compound included the general office, a large clubhouse-restaurant-guest quarters, commissary, hospital, meat market, about 50 first-class employees' residences, tennis courts and a polo field. A nice tropical palm-fringed sandy Atlantic beach was nearby. The native town, separated from the Company compound by a small bridge over a river, was a typical, well-ordered Spanish village with its park, theater, club, a few small family stores, several blocks of residences and "Pepy's" cantina conveniently located just across the bridge.

Most all personnel at the executive level in the offices, in the sugar mill, machine shops, hospital and on the farms were English or Jamaican colonials. They retained some of the old customs like wearing jackets at all meals in the clubhouse — despite the heat! There were only about a dozen Americans all in the accounting and merchandise departments. Well-educated Cubans filled a few executive and all supporting positions in the division.

On the farms, during the *zafra* — cane harvesting season — the cane was cut by hand, hauled by oxcarts to the railroad, loaded into cars and taken to the sugar mill. All cane cutting laborers on the farms were Haitians, brought in by the thousands under contract for the season. They were taken back after the harvest.

My job in the merchandise department was secretary to the

superintendent, Jack Kansas, an American who later became a vice-president of the Company. His assistant was Joe Hanlon, a young Bostonian, who was helpful and friendly.

The manager of the division, Harold Harty, a fine Jamaican gentleman, was an avid polo player and captain of the town team. There were three other teams in the different farm districts. The town team was short one player and Mr. Harty asked me if I cared to try out for it. I had ridden my brother Wilbur's horse only a few times on short rides around Hanover but I accepted the offer.

Mr. Harty gave me two polo ponies and a full-time groom to care for them in the Company stables. However, I had to buy an English leather polo saddle, bridles, boots, spurs, helmet and mallets. It took me over two years to pay for the items. Three days a week after work, Mr. Harty spent an hour or so on the polo field instructing me in the fine points of the game, which I really enjoyed. After only six weeks, I became a regular player on his town team. We had team practice twice a week and the big event was on Sunday afternoons when we would take on one of the farm teams. There was a small grandstand on the field which always had a small formal crowd of 50 to 100 spectators, English and Cuban.

Being a "kid" on a team of older players, I became the favorite of some of the young ladies who would cheer me on whether I scored a goal, or missed, or fell off my horse! Cold drinks were served after each "chukka" and after the game, the players and a few selected guests would gather in Mr. Harty's large house beside the field for an always jolly session of shandygaffs — half cold Cuban Hatuey beer and half Jamaican ginger beer — a most refreshing drink. Often, after the shandygaffs and a change of clothes, the Hartys would invite his team and their wives to return for some cool Scotch highballs and a buffet dinner which might last until the "wee hours."

My boss, Jack Kansas, was secretary of the Company athletic club in Banes and one day he asked me to assist him on a Sunday afternoon in preparing and addressing invitations to a club party. I told him that on Sunday afternoons I was expected to play polo. He told me to show up in his office — or else. I played polo. The next morning Kansas told me I was fired, quite a come-down in my first attempt at a business career after only six months! However, I walked across the street to the manager's office where, in a matter of minutes, I became Mr.

Mark Trafton, Jr., sugar cane overseer and polo player at Banes, Cuba, 1927. (Courtesy of Mark Trafton, Jr.)

Harty's secretary with the title of "in charge of steamship tickets" since he already had a secretary whom he retired a few weeks later. So I was back in business. This move also got me a suite in a very special residential building instead of the room I had before in the "bull pen." Polo had boosted me up a few notches in the Company from secretary to the merchandise superintendent to secretary to the manager of the division.

When I left the States, I sold all my musical instruments except one alto saxaphone which I took to Cuba. This automatically installed me in the small club orchestra — drums, violin and piano — which played for occasional tea dances. Of course, this eliminated me from dancing with the few single English girls, and some of the good-looking younger wives. I solved this problem after a couple of months by swapping my saxaphone with a tailor in the native town for a custom-tailored white linen suit, a "must" for the well-dressed man.

Latin music was fascinating and new to me because it was practically unknown in the United States at that time. Especially enjoyable were the *danzon, paso doble,* and later the *merengue.*

Although I had studied French for three years, Latin for two and Greek for one year, I never studied Spanish so I took a Berlitz crash course in Spanish just before leaving home. This helped some but my first efforts at conversational Spanish in Cuba included a lot of French words which impressed the *señoritas.*

Most of the American guys would spend the evenings and weekends in Pepy's popular cantina across the bridge. I wanted to learn Spanish so I would often go to the local park with some of the Cuban employees. An evening stroll in the park is an old Spanish custom for young and old. The park, a whole city block, was beautifully landscaped. There were benches enclosed by an oval cement walk, the *paseo* where people leisurely strolled and conversed with one another. The custom was for young females to walk in one direction with their chaperones, and the single males in the opposite direction. But if a male were formally introduced and accepted by the chaparone, he could then walk with his *novia* — and no one else! This led to my meeting blond Maria Eugenia de Cardenas, daughter of a local *colono,* the owner of a large cane farm, one of several bordering Company property that sold their cane to the Company.

Over a period of time, this led to my being accepted as a member of the exclusive Banes Club in the native town. Dances there were very formal and white linen suits were mandatory for men. The linen suits were hot and there was no air conditioning except for a few ceiling fans. Strangely, the Cuban *guayabera* shirt was only worn by laborers, but in Honduras and Panama, it was accepted as "full dress" for social events. Although the dances were always too long, one could take it easy, barely moving to the rhythms of the popular *danzones.* All members had a *carnet* — dance card — with the dance numbers listed on it. After requesting and being granted a particular dance with a lady, the male would enter her name on that line on his card, and she would write his name on her *carnet.* Of course, I had to go through all the chaparones, mothers and aunts for permission to ask a *señorita* for a dance and to enter her name on my card. Before long, I also needed Maria Eugenia's permission if I wanted to dance with another girl.

The next move in the social whirl was my acquisition of a reserved *palco,* a box for seating six in the concert hall-movie theater at the park. I had to have seats for the chaparones, also for Maria Eugenia's single sister, Adela, a stunning brunette. One of the concerts there was by the famous Cuban pianist-composer, Ernesto Lecuona, who had taken time out from a European tour to make a few appearances in his own country.

Ever since then, his music has haunted me — I even memorized his difficult *La Comparsa* on the piano. I also learned Spanish.

Transfer to the Farm

After five months in the manager's office, Mr. Harty decided I should transfer to the farms as a timekeeper where I could learn the business and also get an increase in salary from $1,500 to $1,680 a year. There was one condition which he definitely insisted on to the Superintendent of Agriculture — that I be placed on Tasajeras Farm which adjoined the town of Banes so I could remain on the Banes polo team (and not go on a farm polo team!)

Life on a cane farm was another drastic change for me but I liked it and could still enjoy the social pleasures of Banes, with my special quarters still provided there for weekends, and only a 25-minute ride on horseback from the farm to town.

Dressed in white jacket and jodhpurs, black leather riding boots and a white helmet, riding a spirited horse around the cane fields was fun. But the worst part of being a timekeeper was making the payroll every two weeks. Doing the calculations from time sheets and handwriting several large ten-column, 36-line sheets, with carbons, using an indelible pencil, was an ordeal which always ran late into the night. And working in a hot, humid room on a large drafting-type table with a kerosene lamp made matters worse because one smear of a sweaty arm over the indelible writing meant re-doing the whole page. But I persevered and in only three months was made overseer of the farm.

As an overseer, I was assigned a timekeeper, H. Dalzell Piper, about 15 years older than I. He was English, had been with the army in India and after that had been "Keeper of the King's Stud" in England. He was an accomplished horseman, could hypnotize snakes, and cut quite a figure when we would ride into Banes on our horses. He wore a fox-hunting suit with a bowler and had a coiled foxwhip which he would occasionally whip out about 20 feet against the rear end of little kids who would jeer at the strange sight as he rode by. He could also outlast all the gang at Pepy's cantina. This companion was interesting with a never-ending line of hair-raising tales.

His breakfast, which I went along with, was a tall glass of one-third honey, one-third cream and one-third Bacardi Anejo rum. Then, for a bracer while riding the cane farm later in the morning, he would get a boy to climb a coconut palm, cut a green coconut and remove the stem end. Then Piper would pour out half of the coconut water and, reaching into saddlebags, take out a bottle of Bacardi Corriente and fill the half-empty coconut — indeed a cool, refreshing drink. Lunch was our big meal, followed by a siesta, then tea with toast and marmalade before dressing to ride our horses into Banes for the evening.

Piper would hire a Cuban to make up the five-page payroll for him. Obviously, he had access to funds over and above his timekeeper's salary.

When the Haitian men were brought in from neighboring Haiti to cut cane, an occasional female always seem to slip in too. Upon arrival, each one was given a metal wrist band with a number, a pair of blue jeans and a *guampara* — a cane knife — like a machete. They were a sorry lot, almost like animals. If asked his name, each would reply in his *Patua* lingo, *Jose fils,* or Joe's son. All of them were just numbers on the payroll. One day I was leading a group of them from one cane field to another and we passed a foul-smelling, bloated dead pig by the field. In a matter of minutes, it was cut up and taken back by two or three of them to their camp — after sampling it on the spot! I guess slavery in the United States was worse; at least these guys got paid and returned to their country where they would partly starve while hoping they would be lucky enough to come back to Cuba for the next crop to earn some more money.

I had about 150 Haitians on my farm. They lived in long, narrow, open wooden buildings, about 40 in each structure, where they were furnished hammocks. Each camp had a Haitian foreman and a cook. It was eerie listening to them at night as they chanted and danced around a huge bonfire in some voodoo ritual, spurred on by drinking fermented cane juice — *guarapo.*

In the morning, they would be assigned in groups of 40 to cane fields ready for harvest. The large fields, about 25 acres each, were all separatd by wide, clean fire lanes on all sides. Cane fires were terrible on dry, hot and windy days. The cut cane was loaded onto large crawler-type carts pulled by three or more yokes of oxen. The oxmen were Cubans who had their own shacks by the pastures. Each of them knew his oxen "personally" because each ox had a name and would respond to orders hollered to it in loud, uncouth Cuban terms as the driver walked along beside them.

The *zarfa* lasted from four to seven months, depending on climatic conditions and the age of cane fields. During the "dead" season, there was little to do on the farms aside from routine fertilization, some replanting and cleaning firelanes with a few Cuban laborers. Consequently, I had several opportunities to travel on short local trips of a few days each. One of these trips was to the large port city of Santiago, the capital of Oriente Province. There, Teddy Roosevelt's Rough Riders (they were on foot, having left their mounts in the States) attacked and defeated a contingent of Spaniards on San Juan Hill, July 1st, 1898. Cubans called the hill *la loma de las espinfas* — hill of the Springfields. This referred to the U.S. Army issue of Springfield rifles scattered all over the hill after the battle. *Espinfas* was as close as they could get to "Springfield" in Spanish.

Another attraction in Santiago was the Bacardi rum distillery there. Visitors, many from Fruit Company ships and other passenger ships stopping at the port, were given an interesting tour of the whole distilling process, followed by a visit to the lush, beautifully decorated cantina. There were no restrictions on the number of free drinks concocted from the several different Bacardi rums. The guide, an American, who led several tours each day, must have had a stainless steel liver! On one of those visits, I met Emilio Bacardi, son of the owner.

After nine months as overseer, I had finally saved a few dollars and was ready for my first vacation to visit the family in Hanover. So, after saying *adios* to Maria Eugenia, and with a promise to return from vacation in a few weeks, I left on a "sugar boat" for Boston.

While at home, I received a letter from the Boston office notifying me that I was being transferred to Panama as an overseer in the new Chiriqui division where they had just started hewing a banana division out of a tropical rain forest. In addition, they enclosed a railroad pullman ticket from Boston to New Orleans and a steamship ticket direct to Balboa in the Panama Canal Zone — with no stop in Havana on the way down. Also enclosed was a handwritten personal letter from John Mitchell, newly appointed Assistant Manager in Banes, saying that he thought I would have a better chance of getting ahead in the Company by being in the banana end of the business and in a new division. There was also a rather pathetic short note from Mr. Harty to me saying he had considered me almost as a son, and how sad it was to know that I would not be back there. And that he would arrange to sell my polo equipment and have my belongings packed for forwarding to Puerto Armuelles in

Panama. I know there was a broken heart in Banes, and a sad one in South Hanover when I thought of Maria Eugenia. I never knew if the management there explained my transfer to her or if the letters I sent her were ever received. I sailed from New Orleans to Panama on September 5, 1928 to begin another tropical adventure.

Introduction to Panama

Another drastic change in my life was arriving in Cristobal-Colon in the Canal Zone on September 10, 1928. At the Fruit Company office, I was given a ticket for the train on the Panama Canal Railroad to Balboa on the Pacific side of the Isthmus, a small cash advance, and told to report to the office in Panama City. There I met the Company agent, Sr. Tomas Jacome (later best man at my wedding) who checked me into a hotel for a two-day stay waiting for transportation to Puerto Armuelles, up the Pacific coast near the Costa Rican boundary.

My stay in Panama City was not very exciting. Being short of cash, I walked up and down the Avenida Central window shopping and admiring goods displayed from all over the world. "Panama City, the Crossroads of the World," was well-named. One night was pretty wild (but not for me) when about 5,000 sailors in their "summer whites" came on shore leave from several U.S. Navy ships transiting the canal. They literally took over the town, including all the Central Avenue bars and nightclubs, because the Panamanian police had wisely decided to take the night off!

My transportation to Puerto Armuelles was a small, beat-up old coastwise schooner, the *Berta* which chugged along at a snail's pace for four hot days and nights over smooth seas. The captain and crew of three rather young, ragged, dirty guys were surprised that I could understand and speak Spanish so they regaled me with wild seafaring tales of "The Panama Main." The food was terrible except for a few fish they caught trolling. On the second day out I learned that the cargo was a load of dynamite and caps for the contractor building the government railroad from Puerto Armuelles to Progreso. Had I known before I would never have boarded *Berta!*

Approaching our destination I admired the beautiful Secas and Paridas Islands where, years later, I would spend many happy days. The coastline was an unbroken vista of green tropical rain forest, a solid mass which, miles away, faded into a light blue haze over the high mountain range on the horizon.

At Puerto Armuelles a small block of jungle had been cleared and a few labor camps, long narrow buildings on stilts, had been built along with large *bodegas* for storage of equipment and materials, offices, a machine shop, and a small radio station and tower. The government railroad, only a partly finished, narrow strip cut out of the jungle, stretched back several miles into the interior where banana farms were to be planted. Eventually, the line would hook up with the existing railroad from David at Progreso.

There was a small cluster of native thatched palm huts along the beach where some of the local labor lived. The Company had commenced building the large steel wharf where, in a couple of years, stems of bananas would be loaded into ships for west coast markets. When the *Berta* arrived late in the afternoon, another coastwise boat was tied up at the temporary landing where cargo was discharged by a crane, so they rowed me ashore in a small, leaky dinghy which capsized in the light swell along the beach. From there I waded ashore, taking a few Saints' names in vain, to commence my career as a banana cowboy. Fortunately my small wardrobe trunk and suitcase were unloaded later that night at the dock. My arrival had been announced by Morse code from Panama.

I was greeted on the beach by a Company Panamanian lawyer, Licenciado Don Carlos Bieberach, later a justice on the Supreme Court of Panama and a close friend of mine for all the fourteen years I lived in Panama. He had arrived there from Panama City the week before to study engineers' maps of Company land acquisitions in the area. The first-class employees living there bunked four to a room in what would later be labor camps. The temporary division headquarters was in a few old wooden buildings in Progreso, about 25 miles from Puerto Armuelles on the banks of the Chiriqui Viejo River where there had been a small sugar plantation, sugar mill and cattle ranch in the past. Access to Progreso then was by railroad from David, the capital of the Chiriqui Province, a small city about 40 miles distant in the center of a large cattle raising area. Near David the small port of Pedregal accommodated coastwise schooners from Panama City. There was no overland route until the Interamerican Highway was completed in the early 1960's but an airport was built there by the United States Army during World War

II. Eventually Progreso would be connected with the railroad then under construction through the rain forest (later banana farms) from Puerto Armuelles.

My first night in Puerto Armuelles, I was assigned a bunk in a room with two American engineers and a dragline operator. The nightly amusement was a poker game for big stakes which I could only watch. No money! The drinks the players shared with me I thought were awful. It developed they were Bay Rum and water! The Company had not yet been granted a license for sale of liquor. However, a few weeks later hot beer (no ice or refrigeration) came on line and that was the staple drink for over a year except for an occasional bottle of the local Gorgona rum brought in by travellers from David. It wasn't much better than the hair tonic variety!

The schedule for me was to take over planting a "woodland" farm which was just underway in Esperanza but stopped for lack of labor. The native *Chiricanos* from the interior were "cowboys" and had no desire to work in the sweltering rain forest jungles, so the Company had to import labor from Nicaragua. While waiting for a shipload of those "axe and shovel men" to arrive, I was assigned as Assistant Superintendent of the Material and Supply Department in Armuelles, checking arrival and distribution of construction materials and equipment to the different engineering and dragline camps in the jungle. Fortunately, that assignment only lasted for six weeks before a boatload of *Nicas* arrived. About 200 real tough, but hard-working, men who were anxious to work for the relatively high wages in a new banana division as compared to the poverty-stricken situation in Nicaragua.

There was one experience I shall never forget during my short stay in Puerto Armuelles. A wharf stevedore, Dave Spence, and I took a Sunday stroll along an engineers' narrow survey line cut through the jungle from the camp area up into what, once the forest was cleared away, would be the location of the main office, then the hospital, followed by the clubhouse and then the rows of residences for first-class employees. We were well-armed, he had a rifle, a six-gun and a knife and I had a 45 automatic which I had bought from an engineer on my first payday there. As we walked slowly and laboriously along the rough *piquete* — trail — sweating profusely, admiring the plants and occasional birds in the heavy forest, we suddenly "froze." About 15 feet in front of us, an enormous black panther (a rare species even then) had just dropped down from a large tree branch over the trail. The beautiful animal looked us over for a minute or two and then slowly moved away into the heavy woodland understory of low plants and vines. Not a shot was fired — I couldn't even move my arm to reach for the 45 — I felt as though my circulation had stopped. I have never been so scared in my life!

Dave and I just looked at each other, we didn't speak, we didn't have to, as we slowly and silently retraced our steps along the trail back to camp where we took a shower and put on a change of clothes.

The 200 Nicaraguan workers, led by a guide, went by foot through the forest to Esperanza where they constructed two large barracks of poles and thatch which they cut in the jungle — their quarters for the next year or more. I went part of the way to Progreso on a work train and the rest on foot through mud and water along the remaining railroad right-of-way. My belongings went on a tractor supply wagon, operated by Bob Keiswetter, over a circuitous route, first along the beach for a few miles, then up the high banks of the Chiriqui Viejo River to La Esperanza — a two-day trip if it didn't rain.

At Progreso I was given a horse and saddle, a pair of saddlebags with a few emergency rations and water, and told to follow the trail for five miles down the river to Esperanza. The answer to my doubts about following the trail was "the horse knows the way." The horse was no polo pony (I had played polo in Cuba in 1926-28.) but it was experienced in plodding through brush and mud. So, armed with my 45, which I had only shot a few times, I arrived safely in La Esperanza to begin life as a Woodland Banana Farm Overseer.

La Esperanza, *hope* in English, was a clearing along the right-of-way of the future railroad. There were a number of *manaca* shacks for the overseers, timekeepers, and foremen, a stable with about twenty riding mules, a messhall and a medical dispensary, all palm-thatched. The labor camps, as described, were about a mile away toward the river where there was a "commissary." All supplies and building materials arrived once a week on the tractor supply wagon. My *manaca* shack was a new one, finished just a few days before I arrived. Furnishings were a *tijereta*, or large folding canvas cot, with a mosquito "bar" (net), a small folding camp stool, a wooden bench, a galvanized bucket, and a kerosene lamp. I added to those furnishings by acquiring a few empty wooden boxes at the commissary for storage. The shack did have a wooden floor — a few rough hewn planks laid on the bare ground.

Upon arrival I was met by the senior overseer there, Bob Cover, a rough and tough, but kind-hearted bushwacker, a white Jamaican , born on a banana farm in Costa Rica. He too had a "45," but on an ammunition belt with about 24 rounds of bullets. He greeted me with a handshake which almost floored me and invited me into his shack but warned me, "Don't mess around with my woman."

His "woman" was Amy, a young, good-looking and very pleasant black Jamaican, also born in Costa Rica. She was the best *sancocho* (a Panamanian version of chicken stew) cook I have ever known! They had a son, Leslie, who turned out well. Amy stayed at Bob's side for the rest of his life.

Other personnel in Esperanza were: Vernon Selby, dishonorably discharged from West Point for crowning an officer with a chair; Emilio Fabrega, a Panamanian from one of the elite and top political families in the country; Ted Scott, a young Australian wanderer; Red Barkley, just out of Sanford in California; Arnie Halldorson, a second-generation Icelander and an ex-marine from North Dakota; Roy Crosse, a well-educated Jamaican; and Tubby Spence, from California, recently transferred from a banana farm in Costa Rica. These were a mixed bag of guys the Superintendent didn't know what to do with! In the other three districts, it was almost a family affair. Most of the farm overseers, as well as the other executive employees in the division, including the manager, Henry S. "Hank" Blair, a Harvard graduate, had worked together for years in the Almirante-Bocas Division across the mountain on the Atlantic side. These seasoned banana men had just been transferred from that division which was closing down due to Panama disease, *Fusarium oxysporum*, a soil fungus which killed out the Gros Michel commercial variety of banana plants. As Panama disease wiped out a division, it was standard procedure in the early days of the banana business to move "lock, stock, and barrel" to another location of virgin woodland. The move was generally in a different country where an entire new division was developed with the hopes that the operation might last 10 to 20 years, depending on soil susceptibility to Panama disease.

The mess shack where the overseers, except Bob Cover, and timekeepers took our meals, consisted of a long wooden slab table with benches on both sides and a couple of kerosene lanterns hanging above the table. In the *cocina*, or kitchen, to the rear were a large locally fashioned clay oven, an open barbecue-type grill, a large washtub for a sink and a collection of large pots and cast iron frying pans. Water was kept in some diesel oil drums which had been cleaned out. Some badly chipped china and glassware and an assortment of "silver," probably from old Company ships, graced the table. However, in spite of this foreboding description of the dining room, the food, prepared by an old black Jamaican man, was generally excellent and, at times, "Epicurean."

There was no refrigeration of any kind so, save for a few "tins" of imported goodies, everything was "fresh," right out of the jungle or nearby rivers. With hundreds of workers felling the forest every day there was always an abundance of fresh meat, fish, and fruit. The large labor camps had hunters whose only job was to supply their kitchens with food.

Some of the delicacies we feasted on almost every day were: the highly esteemed flesh of *tepescuintle*, a large rodent the size of a small pig; *chancho de monte* — wild hogs; wild pigs (peccaries) that fed in large bands; pigeons; forest deer; chacalaca, pheasant-like birds; crested guan, large turkey-family birds; tinamous, small ground birds; *perdiz* or partridges; wood-quail; many varieties of fish, and crocodile tail meat, secured by dynamiting large pools in the river; bushels of tender hearts of palm from the tall slender palmito palms; fruits such as *zapote, caimito* and, of course, bananas. Rice and beans, brought in by traders from the interior occasionally, were also staples, as was the Company-imported *bacalao,* whole salted and dried codfish fillets direct from Gloucester, Massachusetts. I also suspect that when hunting was not too successful, our stews and soups might have consisted of scarlet macaw, parrot, or toucan breasts and thighs, as well as some monkey meat now and then! The point is, we ate heartily and well. There was one condition before each meal in the evening, the dispensarian, "Doctor" Smith, a black Jamaican, had the power to withhold the single cup of free grog, Gorgona rum, until each diner had downed a couple of ounces of liquid quinine to prevent malaria. I don't know which was worse, the local rum or the quinine, but it was a ritual we all accepted.

Actually the taking of wildlife for food was not as bad as it sounds. As the forests were felled and planted to bananas there would no longer be a habitat to support them and, of course, there remained thousands and thousands of acres of virgin rain forest along the coast after all the farms were planted. With time, in about 15 years, that forest too was destroyed by natives with their roving agriculture and production of charcoal. The railroad, and later a highway, allowed "civilization" to move in and, as elsewhere, that was the end of one of nature's grandest achievements, a pristine tropical rain forest.

A woodland overseer's duties were to give out and price work to contractors for the various jobs — digging drains by hand, underbrushing, lining, digging seed bits, planting, felling, and making roads and bridges. He would check each contract daily, keeping a record of cubications on shovel work in drains and canals as well as hectareas completed of the other field operations. When a contract was completed to his satisfaction he would pass final reports to the foreman who would enter each contractor's work, and the amount due him on the payroll. There were only a dozen or so contractors on a farm and each one paid his own laborers, so payrolls were one-page affairs, quite different from the many pages of individual workers on the sugar cane farms in Cuba. Work days started early taking advantage of the relatively cool mornings when the sun was low. By one or two o'clock in the afternoon most laborers were through for the day and back in camp.

Then, in Esperanza, we overseers would ride our mules to the nearby Chiriqui Viejo River where we would bath and lounge for awhile. The laborers bathed further down the river, but our little *playa*, a gravel beach when the river was low, was always shared with our washwoman, an old *Chiricana* woman, who had a hut close by. As we bathed, she would lay out our clothes she had washed the day before by banging them on a flat rock, and then start on those we had just removed. Her clothesline was some bushes along the river bank. The water in the river was clean and cool since its headwaters were at 11,000 feet up in the nearby mountains. There was a fairly fast flow at Esperanza but the water became saline about eight miles downstream where it entered the Pacific through its large delta.

After bathing and dressing in our "clean" clothes, we routinely rode to the small thatched commissary where we would spend the rest of the afternoon drinking warm beer (no ice or refrigeration, of course). And, after brushing off flies, we would munch on chunks of *bacalao* for *bocas,* or appetizers, ripping pieces from the salted dry fillets in the barrels they had shipped in. There were two kinds of native beer: Balboa and Baru. Our group had their personal preferences for one or the other and when it came to relieving ourselves, we would always use the other guy's empty bottle which we then threw on a pile outside. They would later be collected, put in burlap sacks and eventually returned to the breweries in Panama City. Perhaps they were washed out before being re-filled but we always inspected the contents of the full ones for remains of cockroaches and cigarette butts before drinking out of them.

Planting a banana farm was a novel and interesting experience since I knew nothing about it and had to learn by doing it. Fortunately I had a good foreman, an old black Jamaican, Mr. Gibson, who could make a payroll and had a lifetime of experience on banana farms in the Almirante Division. Bob Cover also willingly shared his expertise with me. And my Nicaraguan contractors and their laborers were a tough and hard working gang, the likes of which I had never seen before or since.

My farm, *Guacimo,* (all of the 40 or so farms in the Chiriqui Division were given Spanish names of local trees), consisted of over a thousand acres of virgin forest and, with my love for nature and the woods, I was really thrilled with the jungle. Engineers had previously cut lines through the whole area and made rough topographic maps to indicate location of primary drainage canals. These lines were then cleared and long, wide, deep main drainage canals were dug by draglines.

Planting started by underbrushing — cutting out all the vines and bush that can be cleared with machetes. Then the area was "lined" using a long bamboo pole to place stakes at the 12 x 12 foot spacings where the bananas would be planted. During these operations, digging of secondary hand drains with shovels got underway. It was back-breaking work using long-handled shovels and throwing each heavy shovel full of clay loam, sometimes from a depth of six or seven feet, up and out of the ditch, clearing a berm on the side by at least three feet. The drains had been staked out and the cubication calculated by engineers.

Planting of "seed bits" followed. Bits, or seed, were pieces of a rhizome with "eyes" (like on potatoes) dug from old banana plants in seedbeds previously planted in the area or shipped in by boat from another division. When planting at the lining stakes was finished, the hard work began — felling enormous trees by hand with axes. Some *espave* trees which were common there were 150 feet tall and had large, wide buttresses. The axemen would build pole scaffolds around the tree, eight or ten feet from the ground above the buttresses to stand on as they swung the axes, well above their heads, for hours on end until the tree toppled. It was sad to see those majestic trees, their branches laden with air plants, ferns, and orchids, crash to the ground, but they would be replaced with the "green gold" of banana plants in a short time.

Sometimes the axemen would line up a big tree with two or three smaller ones to create a domino effect. As the large tree fell, it would break off and take down the others, a labor-saving idea. Try swinging an axe as far as you can reach above your head for six or seven hours a day in the heat and humidity of a tropical rain forest! My shovel and axemen were unbelieveable when sober but unpredictable when drunk, often using a machete to cut up an erstwhile drinking companion!

The banana bits planted in the ground would sprout and grow through the mass of fallen trees and branches. The smaller branches would quickly rot away but some of the thick hardwood trunks would remain for years, unless they were used up for firewood by the workers.

After the forest was felled, section lines and "fruit roads" were chopped and cut through the tangle of tree trunks and branches. Canals and drains were cleared of debris and bridges were constructed. Any misses in the banana planting were supplied with fresh bits so there would be a complete stand of producing plants. Once grown, a banana plant (a perennial herb) keeps sprouting new stalks as the ones with fruit are cut out at harvest. Only extended flooding, fire, drought or disease can kill out plants grown in good soil if they are properly drained, pruned and fertilized.

The post-felling work on a farm was boring for a woodland overseer. With all the forest trees down and the banana plants not yet tall enough to provide shade while riding around the farm, it was hot. However, at that point a woodland overseer is generally transferred, along with his contractors and laborers, to an adjoining jungle area to start work on laying out and planting another farm. Banana overseers, with relatively unskilled laborers, then take over to supervise routine farm maintenance work and almost daily harvest operations when the banana plants commence to bear the large stems of fruit.

The engineers, dragline operators, and overseers who lived in the jungle thought nothing of a seven-day work week. No radios, no entertainment at all except for the payday poker games. The much higher-paid dragline operators and engineers, who had money enough to hold out and last longer, invariably cleaned out us relatively "poor" farm overseers.

But there came a time later on when I tired of losing all my paltry Balboas (the Panamanian currency) every payday. Back in the United States when I was playing on stage with an orchestra in vaudeville theaters, a friendly magician taught me how to undercut a deck of cards. So, lying on my *tijereta* at night, gazing through the mosquito netting at the pink eyes of gekkos crawling upside down along the thatch ceiling, I practiced the "cut" with a deck of cards for hours. Finally, I took Bob Cover into my confidence and told him I could deal him an occasional winning hand if he would split the "take" with me. He was all for it, so in one game, after the big-money boys had polished off a lot of beers, I set up a big full house for Bob. The deck was cut and successfully got it back the way I had it so I could slip Bob's cards off the bottom. Bob kept raising the pot which grew to a big pile of Balboas. It worked! Then, after judiciously letting others win a few, we did it again and wound up with about all the money in the game. There was a lot of hard talk and looks, with hands close to the 45's, but Bob was as tough as they come, and he walked safely back to his rancho.

In the next game a week later, we deliberately lost a little in order to stay alive. Then Bob and I requested a few days off and rode our mules, saddlebags loaded, some 15 miles through cattle country to Alanje where we caught a railroad motorcoach to David, leaving our mules in a corral with an attendant. The small city of David was the capital of Chiriqui Province. We spent two days in a small *pension* and did everything there was to do in town, which wasn't much, but went through all our winnings in a hurry. My only bring-home purchase was a little portable wind-up record player and a few records. I guess I needed a little music in my jungle habitat although the querulous flute-like calls of the small tinamous at dawn were hard to improve on.

Another diversion, although infrequent, was to ride our mules across the river and then about eight miles to Divala, a small village of thatched huts. The rain forest gradually gave way to semi-open plains there and native Chiricanos were small farmers or cowboys on large cattle ranches which supplied the Panama Canal with beef. On special "Saint's Days" they would celebrate at night with a dance in a large *manaca* shack. The orchestra was an accordionist. We were a big attraction and always welcome guests but had to check our sidearms with the *intendente* (a one-man police force) and remove our riding boots and spurs before dancing on the dirt floor with the barefoot Chiricana girls.

They were very short but pleasant and good-looking mestizas with long black hair but with their upper and lower front teeth filed to sharp points. That custom, probably from their tribal Indian forebears, was confined to this rather small part of Panama where it was performed by the mothers, but only on female offspring, once they had acquired their permanent teeth. Of course, a few years later their teeth would begin to decay and fall out and the toothless young women, otherwise healthy and cheerful, certainly lost their charms and good looks when they smiled.

Home-made drinks, fruit juices mixed with cheap native rum, were passed out sparingly at the dance but they were sipped slowly and there was little drunkenness. Of course the *intendente* kept a sharp eye on the dancers and if a man became loud and tipsy he would be tied up to one of the posts supporting the roof in the middle of the room until the party was over. Cruel punishment, watching all the merrymakers dance around him, not one of whom would speak to him! The party was over when the accordionist, after one sip too many, would begin to play some awful sour notes and mixed-up rhythms, making it impossible to dance.

At those dances I often thought of the high-society parties at the Marshall House at York Harbor, Maine, and of the very formal affairs in Cuba at the Banes Club. In Divala I didn't need a carnet to get a dance with one of those *Indias!*

I must include a little story about one of my riding mules, Cantinera. After his six-gun, a woodland overseer's most valued possession was his riding mule. I had three but my favorite was a small, slick, jet black one, very sure-footed on the rough woodland muddy trails and with a nice even gait. Everyone started call her Cantinera (barmaid) because after a few of our afternoon trips to the river, and then back to the commissary (our bar), she refused to go past the cantina so I would have to walk the short distance every day from the commissary to the river and back.

When I first acquired Cantinera she took a dim view of me. Late one hot morning, at a distant part of the farm, I dismounted to check a contractor's work. Ordinarily, with the reins hung over its neck, a good mule would stand in one place until the rider was ready to remount. But this time, as I started toward her to remount, she started walking too, about 100 feet ahead of me, for over one hour along a muddy narrow trail, all the way back to the stable in camp. I was hot and mad. I got into the saddle and rode her as fast as she could go, up and down and around the camp clearing, for a long time until the poor beast stopped, unable to take another step. It took over a week for her to recover. I felt bad about that but she never walked off and left me again.

I guess her revenge was refusing to go past the cantina, probably reasoning that we were both better off there — she tied up and nibbling on sugar cane (remnants of an old plantation) and me grazing on *bacalao* and hot beer!

Ted Edward Scott was one of many characters who appeared in Chiriqui in the early days looking for work. He was a well-educated Australian adventurer seeking his fortune and was assigned as a timekeeper on the farm next to mine in Esperanza. We lived in the same small compound of *manaca* shacks. He interested me with his tales of having been a novice prizefighter back "down under" and of his experiences travelling half-way around the world before arriving in Panama. He didn't stay long in our camp because as soon as he had saved a little money he took off for Panama City.

I didn't hear of him again until some four or five years later when his name appeared as an assistant editor and columnist for the prestigious *Panama American* newspaper. A few years later he was in jail in Panama City for articles he had written about Arnulfo Arias, an off-and-on presidential candidate and later president for a short time, accusing him of Nazi tendencies (which was true). After a few months another newspaper reporter wrote an article, accompanied by a picture, about a jail cell in Panama City where Scott had been detained. Scott had written in large letters on the cell wall "*Arnulfo Arias es un hijo de puta.*" Not long afterward, Scott was out of jail and Arnulfo was in, just before he was expatriated, and he wound up in the same cell that Scott had been in. The picture was of Scott's thoughts about Arnulfo, only now, Arias had crossed out his name and had written above it "Ted Scott."

Ted continued with the paper for some time but he became more and more controversial due to criticism of Panamanian politics — which was not good! After another couple of years, Arnulfo was back in Panama running again for President, and Scott was out, looking for a newspaper job in some other country.

Many years later, in 1954, a long-time friend of mine, Frank Keany, then assistant manager in Banes, Cuba, sent me a clipping from the *Havana Post.* It was by Edward Scott, now a regular columnist for that paper. I quote parts of his column "Interesting if True" for August 11, 1954.

"Bulletin by Jungle Wireless." "Sometimes I have referred in this section to incidents in the jungles of Central America and even while I was writing, I knew that there would be some, or many, who would suspect that I was drawing the long bow. When I have spoken about nine-foot jaguars, or *tigres* as they are known South of the Border, there has been a raising of eyebrows and a mental reflection on the good which could result from the taking of tonic without what goes with it like mint goes with julep."

"In a discussion I had with Francis X. Keany, assistant manager of the United Fruit Sugar Company at Banes, we were talking about old times and old-timers. The name of Mark Trafton, Jr. was introduced. I said I had known Mark Trafton in Chiriqui on the border of Panama and Costa Rica back in 1928, when he was a fresh-faced youngster. Now he is Assistant to Norwood C. Thornton, Director of Tropical Research."

"I well remembered Mark Trafton, with his banana cowboy hat and six-gun. We were about the same age and became quite clubby after we had been introduced in the presence of a case of hot beer and a couple of pounds of *bacalao* and crackers. 'Yes,' I said, waving warmly reminiscent, 'Mark is really a nice bloke. He came straight to Chiriqui, Panama, from school, but he liked the bush.'"

"There was a pause while Mr. Keany made up his mind whether it would be proper to contradict me or flatly or merely outflank me. 'I think he was here in Cuba before that,' he finally said. A few minutes later he had a card in his hand which . . . established that Mark Trafton, Jr., had been employed in Banes, Cuba, as a stenographer on November 13, 1926, and furthermore, he became a full-fledged overseer before he was transferred to Chiriqui on September 5, 1928."

The Railroad Arrives

Just as I was finishing the planting of Guacimo Farm in late 1929, the railroad arrived in La Esperanza and that was the end of the "good life" on a woodland farm. Work trains arrived daily loaded with construction materials and workers. Soon there was a wooden, eight-room bachelor quarters building with messhall, kitchen, running water, toilets, showers, and electric lights. And a commissary with a large ice box. Ice was delivered three times a week from the ice plant in Armuelles, but it took us a couple of weeks before we could stomach the cold beer — it would bloat us up terribly! The Company now had a license to import and sell liquor and an immediate improvement was that we now got a slug of good Scotch whiskey instead of local rum if we downed the liquid quinine at the supper table. Our menu slowly changed from fresh local fish and meats to canned stuff and cold storage items from California. Permanent wooden labor camps were being constructed and work started on houses for the superintendent and three overseers. Esperanza became a district headquarters and included five other farms in addition to those adjoining the center. Additional labor camps and overseer's houses were built on each outlying farm.

In Esperanza there were two new arrivals, a district superintendent, another transfer after many years in the Almirante division, and Dr. Vining C. Dunlap. "Doc," a Ph.D. in Agronomy, from Bowdoin, Maine, had spent several years in Africa and in tropical America looking for banana lands for the Company. Chiriqui was one of his recommendations and it was his opinion, given to the president of the Company in 1926, that the soils in this region would produce good bananas for many years. It is still producing, and is one of the Company's largest divisions. Some of the older divisions died out after 10 or 15 years. Doc worked directly for the president of the Company and now was assigned to check the soils in the jungles of the Golfito area in Costa Rica, which adjoined Chiriqui Province in Panama.

Esperanza would be his base camp and he was provided with two camp cars: living quarters and a kitchen. Camp cars were like mobile homes but constructed on a railroad flat car. His cars were spotted on a railroad siding across from the new bachelor quarters. He immediately became a member of our afternoon beer and *bacalao* club and the principal raconteur with his vast knowledge of and experiences in the Tropics. He and I soon became close friends, both New Englanders, and with similar interests in natural history.

Doc had a cook and housekeeper, Margarita, a good-looking Costa Rican girl, an excellent horsewoman and a never-miss shot with her Colt 38 Special six-gun. While Doc was away on extended trips making the soil survey, she cared for and protected his camp and a little vegetable garden which she tended. The whole area was enclosed with a high chain-link fence and locked gate and was off-limits to the Esperanza crowd — all except me! Margarita had a record player too and had heard a couple of records echoing from my *manaca* shack. One evening at our beer session Doc asked me to bring my records over to his camp and have dinner. We listened to my records and theirs while sipping several Dewards White Label Scotch highballs and then feasting on tamales. (White Label became *the* drink for United Fruit employees for the next 40 years that I know of, perhaps longer. At first the cost in the commissary was about $1.50 a fifth!) That evening was the beginning of a lifelong, almost a family, relationship between Doc and me that lasted until he passed away 40 years later in New Orleans. He inspired me, taught and guided me directly or indirectly, for the remainder of my years with the Company.

Doc's specialty was soils but he knew the names, scientific and local, of almost every tree in the forest, as well as many of the orchids, birds, and animals. He became Director of Tropical Research for the United Fruit Company and I worked for him in Honduras for many years. He was probably the best-known employee in the Tropics, highly respected by everyone, and a guiding influence in several critical decisions which literally saved the Company in later years.

The Transition

I have mentioned that a district superintendent had been assigned to Esperanza to supervise the five farms and make preparations for harvesting and shipping bananas which were now commencing to "shoot" the flower bud. Potential banana production, barring blowdowns or floods, was around 625 bunches per acre yearly. With bunch weights averaging 80 pounds, this was 25 tons of bunch fruit per acre. As bananas are now packaged, this would be about 1,000 boxes per acre.

A district clerk was also assigned to Esperanza to handle the office work and records. Up until the superintendent arrived, we woodland overseers answered only to the Superintendent of Agriculture whom we rarely saw since he lived in Progreso. He only came to my farm once to spend a few hours riding around checking the general quality of the work. A young guy from the accounting department in Puerto Armuelles also appeared once to check some of the field work against my payroll. He didn't hang around long and was scared to death of the sound of howler monkeys which I told him were tigers. And he could hardly walk after a couple of hours in the saddle of a rough-gaited mule, provided especially for him!

Life in Esperanza was different now with the railroad, the new buildings and people coming on motorcars from port almost daily to check this and that but mainly just to see what a banana farm looked like. We could still see and hear the flocks of macaws and parrots flying over at daybreak, and the laughing falcons, *guas,* calling at dusk. But with so much activity and so many people around, the jaguars, ocelots, and pumas had retired far into the remaining forests in back of the banana farms.

The 5,000-acre district was now planted but in the three other districts, where about 10,000 acres were in bananas, there were still some 20,000 more acres of jungle to plant. Woodland overseers and their specialized labor were at a premium and I was lined up and looking forward to starting another new farm near the Costa Rican border. In the enjoyable solitude of my jungle life, I wasn't even aware of the Wall Street stock market collapse in October, 1929, which was the start of the great depression of the 1930's. However, its effects were beginning to be felt by the United Fruit Company — and very much so later on. In addition, another division, Puerto Castilla in Honduras, was closing down due to Panama disease so it was important to get the Chiriqui Division in production and shipping bananas.

Some experienced banana farm overseers and timekeepers were transferred from Castilla to run the newly planted farms. Woodland labor didn't go for ordinary farm work — pruning bananas and harvesting the bunches paid far less than their shovel or axe contract work. But by now some of the local Chiricano labor from the interior were coming in for the steady jobs on the farms where living conditions were good, the work didn't require too much physical effort, and a payday was assured every two weeks.

To get banana production under way, the former Superintendent of Agriculture was retired and the Superintendent from the Tela Railroad Company in Honduras, Ed M. Cobb, was temporarily transferred to Armuelles. He was a tough *hombre* and the top banana production man in the Company. His first move was to visit all districts and personally talk with each first-class employee. The day he came to Esperanza, the last district he visited, we were all concerned, to say the least, because in the other districts he had summarily fired several of the old employees who had come from Almirante. He decided to start work on only one woodland farm, in the Blanco District close to Costa Rica, which he assigned to Bob Cover. That made me happy, and Bob took most of my woodland contractors and labor with him to the new job and that also pleased me. But I was worried about my future as Cobb had retired the district superintendent and one overseer in our group. He hadn't spoken to me, but when ready to leave, he told me to pack a suitcase and go back to Puerto Armuelles with him. I wondered what was up — being taken for a ride and getting fired in a big way, or what? Riding the motorcar back to Armuelles he told me he had looked at my record and that I was the only one in the department who had any administrative ability and that I would be the Chief Clerk of the Agricultural Department and his right-hand man! Arriving in the office, he told the clerk there that he was transferred to a district office. Then he told me to make out a 1672 (a personnel change form) to give me a raise from $1,800 to $2,100 a year with the title of chief clerk and to set up an efficient office system and staff.

That was on May 7, 1930, and it was my job for the next three years. Cobb had already reserved one of the best rooms in a brand new bachelor quarters for me.

Then he invited me to his new house, the best one in town aside from the manager's, where we had some White Label highballs and a big supper, as we carefully exposed bits of our personal past to each other. Cobb later married Gladys Evans in Tela and was the father of Ginger and Maggie. Later they were divorced and Gladys married Floyd "Oscar" Blackmon. They became close friends of our family.

That weekend I went back to Esperanza to pick up the rest of my gear and attend a big party which my number one Nicaraguan contractor, Tomas Mairena, had for me in his large *manaca* camp — with about seventy of his laborers and the other Esperanza overseers. I rode Cantinera to his camp with no difficulty since the old commissary along the way had just been demolished! I was presented with a *cruzeta*, a long narrow type of machete but with a straight blade like a sword and with a brass grip with the guards extending about two inches on each side, forming a cross — hence the *cruzeta*. It was a prized possession of all Nicaraguan "gladiators" who could afford one. What a night with many guitar and vocal numbers, all going on at once and gradually becoming more dissonant. There were a few fights, but no major blood letting. I was feeling bad from eating and drinking so much but they strung up a hammock for me and after sleeping a couple of hours, I recovered enough to ride Cantinera back to our camp in the morning. Then I said *adios* to the rest of the gang there (although I would see most of them off and on later) and took a motorcar back to Puerto Armuelles.

Puerto Armuelles

And that was the end of a rather short (18 months) but exciting and enjoyable chapter in my checkered career — that of a banana woodland overseer.

By this time, Puerto Armuelles had become the division headquarters and, like all Company towns, it included most of the amenities for a comfortable home and family life for the employees living there. The wharf had been completed and shiploads of construction materials arrived one after another. A nearby river had been dammed for water supply and pumped to a large standpipe. Over a hundred labor houses had been constructed along the beach for about a mile, from the wharf and native village at one end, to the ocean-side jungle at the other end. This huge labor camp would be known as Silver City later on. In back of the labor camps there was a large machine shop, power plant, material department building and railroad yards for storage of rolling stock. Then a long airstrip was built which also served as a fairly good nine-hole golf course.

At one end of the golf course was the large merchandise department building with its ice plant and cold storage facilities, and at the other end, the jungle. Immediately in back of the airfield was a line of buildings, a radio shack and tower, the two-story general office building, a three-story hospital, the manager's home, a large messhall-club building, tennis courts, (later a private club-bowling alley building) and nine eight-room bachelor quarters, each with a large central sitting room. Con-

View of Puerto Armuelles, Panama, before the 1934 earthquake destroyed the original wharf shown here. Banana pioneers landed on these beaches in 1927 and converted the wilderness into a large Company town in a few short years. (Collection of Clyde Stephens)

This United Fruit Company hospital in Puerto Armuelles, Panama, was built in 1932 and served thousands of Chiriqui Land Company employees and the community. The building was razed in 1970 and replaced by a new government hospital built across the tracks. (Courtesy of Tam Syme)

tinuing away from the ocean and up a hill there were about thirty widely-spaced houses for first-class married employees, a small school house, a swimming pool, and, atop the hill, the tall water storage tank. All buildings were constructed of California redwood because of its resistance to termites.

All the camps and houses were built on "stilts" about seven feet off the ground to provide space beneath them for cooking in the labor camps and for laundry rooms and storage facilities in the family houses. A church was soon constructed with private funds and there was a cemetery on another hill just outside of town near the railroad line. All of this construction was accomplished in a little over a year and it continued at a rapid pace after that for several more years. And there was a lot of construction on the farms as new plantations opened up, more supervisory personnel and labor were brought in, and banana shipments commenced to west coast ports.

In laying out the town area, many of the original forest trees were left — an inspired landscaping idea. The clearing for the town was a half-moon shape about one mile long, the Pacific Ocean bordering one side and the pristine rain forest jungle in the back. It was located near the basal end of the Burica Peninsula which created the anomolous situation where the sun would *rise* over the Pacific Ocean! From the very first, thousands of ornamental bushes, plants, and grafted fruit trees were imported from Summit Gardens in the Panama Canal Zone and planted along the streets, in the yards, and around the houses in Armuelles and also on all of the banana farm housing areas.

Up the coast, and in back of the town, the terrain soon became a series of heavily forested high foothills, cut by deep gorges of small rivers flowing down to the ocean. The adjoining great block of thousands of acres of rain forest continued into Costa Rica to the northwest; to the Cordillera de Talamanca mountain range to the north with its 11,400-foot volcanic peak, El Baru, and to the level flood plains of the Colorado and Chiriqui Viejo Rivers to the northwest where the banana farms were planted on the rich deep clay loam alluvial soils. The railroad from Armuelles to the banana farms, and on to the city of David, cut through the forested hills and valleys to the northeast for about ten miles before emerging into the nearest banana district. Then there was some 40 miles of railroad loops and loading spurs in the banana area, followed by about 30 miles more through semi-open pasture and cattle country to Concepion and on to David.

The Staff

At first, and for a long time, most of the top brass were North Americans, many with families, having been transferred from other divisions. They filled the many administrative and clerical positions in the various departments in port: executive, agricultural, accounting, engineering, shipping, railroad, mechanical, materials, medical, and merchandise. The legal staff consisted of several highly-qualified Panamanian lawyers.

Mr. Cobb and I had plenty of space in the large two-room agricultural section of the general office building. My staff consisted of a secretary, Lilia Jensen (a beautiful but very shy girl, daughter of the chief engineer, a Dane, and his Panamanian wife), "Captain" Edwards (an Englishman who couldn't handle his liquor, formerly accountant for the Singer Sewing Machine Company in Buenos Aires, Argentina, and John Tower (a "hurried" transfer from the Company accounting department in Santa Marta, Colombia, after a gun fight). Surprisingly, they were all excellent workers. How we would have enjoyed the small digital computers in use today instead of the endless hand-cranking and carriage-shifting of the heavy and noisy mechanical adding machines in use then.

Now that I had the office set up, Cobb took me with him several times a week on motorcar trips to different districts where we would ride a banana farm on mules with the district superintendent and the overseer, an all-day job. I also accompanied him on some fruit cuts. So I learned the banana farm business fast, and from an expert. It was a tough school though. I was often emotionally upset by Cobb's direct, brusque and often inconsiderate treatment of some employees, but that was his job, "to clean house," and he did it. He soon brought in several of his top overseers and Kenneth B. Block, a district superintendent from Honduras. KB would become Superintendent of Agriculture in Chiriqui so that Cobb could return to Honduras. KB and family became close friends of ours for my remaining eleven years in Armuelles, and for many years thereafter until my retirement.

Cobb told me to teach KB the tricks I had learned from him so when Cobb left, my routine was the same as it had been. I was like an *aide-de-camp* who knows the whole procedure, and was often directly responsible for the results, but didn't get the top pay!

The division manager, Henry S. Blair, a gentleman of the old school, had been manager of the Almirante Division and was transferred to Chiriqui in 1927. With Cobb "carrying on" in the division without ever consulting him, Hank was worried about his job, even though he was at the top. He often entertained me with cocktails in his house, along with his retinue — Jerry Baron and others he had brought along from Almirante. That was how they kept abreast of what Cobb was up to. As time went on, Hank Blair and I became very good friends. In fact, I practically ran his job for him later on as you will learn in another chapter. K. B. Block, the Superintendent of Agriculture, also became a good friend of Blair and was a great help to him over the years he remained in Armuelles.

Life in Puerto Armuelles

I missed the exciting life on the woodland farm but with the untouched jungles so close to town, I could still enjoy the pleasures of nature with weekend walks into the forest. Just looking out of a second-story window at one end of the general office building, and about 40 feet away, in a large tree on the border of the jungle, I could see a giant sloth. The animal spent several years there and was a rare sight with its slow movements and its upside-down posture and with much of its long fur covered with a growth of green algae. Just the sight of it effectively kept most of the townfolk out of the forest! Bands of howler monkeys were heard and seen daily and, of course, great flocks of several species of parrots. The beautiful scarlet macaws, giving their raucous calls, were common sights flying over at dawn and at sunset. And I admired the hundreds of forest birds, the many flycatchers, antbirds, hawks and the brightly-colored trogans, motmots and toucans as well as the brilliant butterflies. All of the large animals — tapirs *(dantos), tigres,* mountain lions, and bands of peccaries were close by and seen frequently by the few daring individuals who would venture into the jungle. But I was especially attracted to the great variety of orchids and the ease with which some of them could be collected. That became one of my major hobbies during all the years we spent in the Tropics. I commenced by sending a few dried specimens to the Director of the Summit Gardens in the Canal Zone requesting the favor of identification, and a source of literature for reference. The director, who I did not know at the time, was Paul H. Allen, and there will be more about him later.

I had played golf with my dad a few times at the Marshfield Country Club before leaving the States but was no pro by any means. KB was interested in golf so we both played the nine-hole course almost every day after work. Among other regular partners, or opponents, were Jerry Baron (port superintendent), Sid "Slim" Banack (railroad master), Casey Jones (railroad conductor), Rudolf Jensen (chief engineer), Lane Evans (farm district superintendent) and Jose Mase, the chief draftsman. There was usually some heavy betting during the game — and afterward in the clubhouse as we rolled dice for drinks. I practiced a lot and became good enough to hold my own most of the time and win some money now and then. I played golf several times a week for the remaining 11 years I lived in Panama. Mr. Block became a top-ranked golfer and participated in several professional tournaments in Central America during the next 30 years.

The first tee was close to the office and the ninth hole was near the clubhouse. For us golfers, the game became a ritual — exercise, a chance to make a little money, and a way to get to the club and sit around rolling dice for the inexpensive (20 cents) Scotch highballs. Another golfer, Mike Reagan, the accountant, later told the story that it took him five years to discover that he could get to the club from the office by walking along a concrete sidewalk for four minutes instead of playing nine holes of golf! The other guys in the accounting department, mostly Boston Irish, and good at drinking, knew the short way from the start and never did take up golf. The dice games, with a box of five poker dice, were exciting — Chingona, Liars Dice, and Beat It You Bastard. The time passed quickly until the final messhall call for supper — sometime later. I never could get interested in tennis — too hot!

The messhall was operated for the Company by a Chinese, Ling Yung, latinized to Leon, a smart and friendly person who also ran the bar for his own account. At Christmas he would give his "best customers" beautiful gifts of Chinese wines and an assortment of rare food delicacies. Everyone had credit and signed bar chits like there was no tomorrow — until payday arrived — with no balance! Poor Leon, not the Company, would hold the bag for another month. He ran an excellent messhall, having been in charge of an army mess in the Canal Zone before the Company hired him. The food was large helpings of American-style "meat and potatoes," but with hot *jalapeño* peppers mixed in!

In the adjoining native village there was a pit for cockfights, a major Panamanian sport for young and old, rich and poor. A chance to win, or lose, a good rooster or a lot of Balboas and, sometimes, lose a little blood if you disagreed with another owner over how he tied the spurs on his rooster!

The pool was popular with the married couples and their small children. The beach was a blue-black fine volcanic sand from the Baru volcano and, with a 12-foot tide, was not suitable for bathing — a factor previously considered when the labor camps were constructed close to the ocean.

There were black-and-white movies in the clubhouse on Saturday nights, and dances on holidays, of which there were many between the American official ones and the Panamanian "Saints' Days." A special train made the circuit of all the farms to bring the "farmers" in and to take them home after the movies or dances. Performing the long and fast Panamanian dance, *La Pollerra*, was really "hot stuff" with only a few ceiling fans which were not enough to cool down the exuberant dancers. Late one night, KB dragged a large washtub full of ice from the bar. He sat in a chair in the middle of the hall, cooled his feet and legs in the tub — while cooling his inner man with highballs for a couple of hours — as dancers circled around him.

In the native part of town "across the railroad tracks," an enterprising Panamanian built a large two-story motel-type building, the Miramar, with a large bar and lounge and a separate restaurant on the first floor and about a dozen rooms upstairs. Occasionally a salesman would rent a room for a couple of days but most were generally occupied by unattached native girls. They were combination waitresses, saleswomen, and social companions who received a cut on all the drinks they sold to bar customers — mostly single male Company employees from the accounting department and sailors whenever a ship was tied up at the wharf.

At times the crowd would get a little boisterous and one night I happened to be there (I was not a regular) when the local *intendente* sent his squad of six soldiers armed with loaded rifles to raid the place. They arrested all the men, mostly Company employees for "disturbing the peace" and marched us to the *calabozo,* a large room in the combined Port Captain, customs and police building by the wharf. Previously Jerry Baron had introduced me to the Port Captain and the *intendente* since I was practically the only *gringo* in town who could speak Spanish. So we were all seated in the *intendente's* office awaiting the sentencing — each one giving his life history to the clerk who laboriously wrote it out in longhand. All were to be fined five Balboas and spend a day in the clink. I was the last one called up and the *intendente* winked at me and told the clerk in a stern, official voice, "*Libertad especial para el señor Marcos.*" What a howl went up from behind the barred door as I walked out! However, I went directly to the office and had the night shift telephone operator call the local legal representative of the Company. About two hours later, around four in the morning, all the rest were set free and I became a hero!

The restaurant in the Miramar was a good one, operated by a Chinese, George Ng, who was the cook, and served excellent Chinese food. It was a popular place for Company married employees to "eat out" and also with the bachelors who frequently enjoyed a change from Leon's American-style hash in the Company messhall.

Deep sea fishing was wonderful, out toward the Paridas Islands or off the peninsula to Charco Azul, where the coast became very rocky, and on to Burica Point and around it into the Golfo Dulce in Costa Rica. Catches of *papagallo,* red snapper, Spanish mackerel, kingfish, dolphin, rockfish, jack, tuna, and many others were common. There were also sailfish and blue marlin. The Company had a large converted submarine chaser, the *Cricamola* and a smaller launch used for tying vessels up at the wharf but they were not used for fishing except for visiting top brass. However, Jerry Baron, who was also the official port pilot (with a captain's license for "any vessel, any ocean") would sometimes have to "try out the engines" on the *Cricamola* and would ask a few of us golfing pals to go along. We might just happen to drop a hand line with a trolling spoon over the side during the "trial run" which might take most of the day. Jerry's wharf foreman, "Big Mac," a Jamaican, would also "try out the bar" by serving ice cold Scotch highballs.

Johnny Mase, like me, a naturalist at heart, often went along with me on woodland walks. He had been transferred from the Castilla Division in Honduras where, almost for a song at an auction he had acquired a 32-foot motor-sailor, a large, sturdy, wooden boat made in Denmark. It had been sailed across the Atlantic but its adventurous owners went broke in Panama and auctioned it off to get money to return to their homeland. On his first vacation, Mase went to Cristobal and sailed the *Whiskey II* through the canal and along the Pacific coast to Armuelles. I spent many hours helping him overhaul it, repair the engine, sailing gear, and sails while it was "hauled out" for about a year. I eventually became a part owner and we had many wonderful experiences "sailing the Pacific."

In the summer of 1932, with savings of about $300, I went on vacation, flying to Panama City and taking a Company ship from Cristobal, via Kingston, Jamaica, where it loaded bananas, to New York. The ship was filled up with paying passengers and the only space for a Company freeloader was a bunk in the medical emergency room, but it served the purpose.

The passengers, who had boarded in New York, were mostly young female school teachers or store clerks on their summer vacation. Of course, before the ship got to Cristobal, its officers had the pick of the crop but there were still quite a few left looking for a guy to buy them drinks.

After arrival in the United States, I rode the New Haven Railroad "owl sleeper express" from New York to Boston, midnight to 7 a.m., and spent three weeks with my family and visiting many of the scenes of my youthful adventures. How places, and people, had changed in just the four years since my last vacation in the States. The great depression was really setting in then, all over the States — and with me, after the boat trip up! But I got a small advance from the Boston office to get me back to Panama.

Catching up with a little Company history, in 1930, the United Fruit Company bought out the Cuyamel Fruit Company from Sam Zemurray for $31,500,000 in Company stock, making Sam the largest stockholder. He then retired. Cuyamel had been a small competitor in Honduras and owned most of the land across the Ulua River from that of our Tela Railroad Company. The Cuyamel headquarters was in La Lima and its wharf in Puerto Cortes. So now our Company had three ports in Honduras — Cortes, Tela, and Castilla. The Standard Fruit Company (a competitor) wharf was at the port of La Ceiba and their banana farms were along the coast and in the Aguan River valley.

The great depression had taken its toll on the UFCO and the value of its stock had declined drastically. Zemurray didn't like that so he went to Boston in 1932 and confronted the board of directors. I quote from Tom McCann's book *An American Company — The Tragedy of United Fruit Company.** (I knew Tom

*With permission from Crown Publishers, Inc.

very well.) "He was greeted, frostily at best, by a board of directors which included Daniel Gould Wing, then chairman of the powerful First National Bank of Boston; former Massachusetts governor Channing H. Cox; (both Wing and Cox were direct descendants of American presidential families, the Jeffersons and the Coolidges); the Lee half of Boston's most prestigious investment firm, Lee Higginson; and Bradley Webster Palmer, a leading Boston lawyer. Zemurray presented an incisive review of the Company's mismanagement, backed by a bill of particulars which included the fact that the stock of the United Fruit Company had declined by almost 90 percent since he had become a shareholder. He demanded to know what the board intended to do."

"Daniel Gould Wing responded for the other directors. He smiled thinly and in a reference to Zemurray's accent, said, 'Unfortunately, Mr. Zemurray, I can't understand a word you say.' Zemurray just looked at Wing for a few seconds and then at the other Brahmin faces around the room. The story goes that he muttered something under his breath as he left. He returned shortly with his hands full of enough proxies to make translation unnecessary. Slapping them down on the long oval table at 1 Federal Street, and speaking very clearly to avoid being misunderstood, Zemurray said, 'You gentlemen have been fucking up this business long enough. I'm going to straighten it out.' Wing and his clique were dumped from the board and Zemurray was back in bananas. He dealt effectively with all problems for the next twenty years and repeatedly returned to the Tropics where he could see the real United Fruit of men and machetes and machinery and vast plantations close at hand. He never confused that United Fruit with the Company in Boston, the United Fruit of paper and proxies, meetings and numbers."

That major shift in top management, together with the adverse effect the depression had on overall operations, brought about some drastic changes in all tropical divisions in 1932. All salaries were reduced 10 percent. Production in the Chiriqui division was stopped for several months and many employees were released, not only on the farms but in all departments. Some of my old pals were out but a few remained as "caretakers" on the farms. Bob Cover kept his position as an assistant district superintendent, at the reduced salary rate. One who definitely benefited by the change was Doc Dunlap. Zemurray knew about him and made him Director of Tropical Research and transferred him to the Company's new tropical headquarters in La Lima, Honduras. Sam named Walter E. Turnbull, his former top man in the Cuyamel Fruit Company, as General Manager of all tropical divisions. Naturally, all of the Cuyamel officers were assigned to important jobs in the different tropical divisions. Armuelles, being partially closed down, was not affected too much by those top personnel changes for several years.

Since administrative office work had been greatly reduced, and due to my knowledge of farm operations, Mr. Block appointed me as Farm Inspector in October 1933.

I still lived in Port and was assigned a motorcar to make daily trips to farms where I would ride with the "caretaker" of each farm, mainly to check on the vast system of drains and canals over the 50,000 acre-project. Drainage had to be maintained at all cost since flooding or high water tables would quickly kill out the banana plants. I decided which canals should be repassed with draglines, which hand drains should be regraded and which bridges and culverts should be maintained. Then, after approval by KB, I would authorize the necessary work on a priority basis.

This slow-down in divisional operations didn't last long as there was soon a banana shortage and things began to pick up slowly. New farm timekeepers were hired by the personnel department in Boston — most of them recent Texas A & M graduates with some knowledge of agriculture and Spanish. All had been ROTC students and were a fine bunch of boys who adapted easily to farm life.

By the middle of 1934 operations were almost back to normal when, at 5 a.m. on July 21, a major earthquake rocked the whole area which was very close to the San Andreas fault. Two-thirds of the wharf and three large overhead powered gantry banana conveyors were gone! The raised buildings in the division had only minimum cross bracing underneath so many slid sideways off the stilts, dropping to the ground. Fortunately there were no serious injuries, and no fires, since almost everyone was still in bed and the kerosene stoves in the labor camps had not been lit. The accountant, without looking, dashed out of his front porch door, but the house had slid ten feet away from the steps so he fell about eight feet to the ground — but no broken bones. My room in the bachelor quarters fell about four feet while the building slid to one side. Everything in the room — the bed, me, washbowl, dresser, chairs, and table — were all piled up in one corner, with water from a broken pipe squirting over everything. I dug out some clothes and made a hasty exit to join everyone else in town looking over the battered remains while in a state of shock. Fortunately some of the camps and residences were still "livable" and the large buildings such as the general office, hospital, club, and nurses' quarters which were built close to or on the ground were not damaged. By coincidence, a converted yacht, the *Nokomis* assigned to the U.S. Geological Survey Service had been working in the area for some time and was anchored close to Paridas Island off Armuelles, almost directly over the epicenter of the quake, so it was well-recorded!

There was a continuous series of somewhat less violent shakes for a couple of months afterward, some still strong enough to break up some of the remaining glass and chinaware in the houses. On the golf course, when one of the minor *temblores* would occur, the ground would break into a series of closely spaced "waves" a few inches high making it impossible for one to stand erect and producing an extremely "queasy" feeling in the stomach!

Late in the afternoon on the day of the quake, KB said that Mr. Blair needed me badly in his office. There was no assistant manager in the division and Blair's secretary, a young married girl, had quit because she was scared — and pregnant. Again, I was called upon to change jobs on a few hours notice, this time as secretary to the manager, a change that lasted for five years. Mr. Blair was very upset with all the radiograms, many in code, coming in by the hundreds, all wanting immediate detailed replies of damage, cancelling fruit cuts, ships' schedules, etc. Employees worked day and night to determine extent of damage to buildings, the railroad, bridges, waterlines and towers and to the wharf. And estimates had to be made of materials needed and cost to repair everything and get back in business. I became a combination manager, secretary, cable clerk, coordinator between departments, and soothsayer to some very worried co-workers. All were afraid the division could close down for good and that they would be released.

The abrupt change from being shaken out of bed by an earthquake in the morning to working for the manager in the afternoon had its rewards however. I got an immediate salary raise to $2,400 a year and was assigned one of the best rooms in town in one of the few undamaged buildings, the single girl's quarters! A telephone was installed in my room the next day. I had a typewriter in the room and a set of the Company radio code books since I was on call 24 hours a day, handling calls at night by messenger from or to the radio station and different department heads. Between calls at night, I was entertained in the parlor by females with coffee, cookies and plenty of gossip. They were thrilled having a man living in their midst and I was a regular at the nurses and telephone operators night shift changes.

This hectic existence went on for several months. Engineers from Boston and other divisions came in to draw up plans for a new wharf and conveyors. Construction personnel from other divisions were brought in to supervise jacking up the fallen buildings and, as a future safeguard, install double cross-bracing under all of them. Gangs of railroad maintenance men worked at straightening the railroad lines and repairing bridges. Electricians and plumbers were everywhere. Jerry Baron worked up a plan to load bunches of bananas by hand onto lighters (barges) and into ships. The lighters were rented and towed from the Canal Zone and, after a few months, ships were being loaded again, slowly to be sure. Thousands of letters, in addition to radiograms, were received and written in the manager's office to coordinate all of this reconstruction and, worse, they all had to be filed.

The lighter-barge system of loading ships by hand was used for about two years until a new wharf and conveyors had been built somewhat south of the old wharf. All planning and reconstruction work, even down to deep-sea divers, was done by Company personnel — local and transfers from other divisions in the Tropics and from the States. All material was ordered and supplied through the Company's purchasing division, the Maritrop Trading Corporation in New York City, an example of how large and self-contained the United Fruit Company, one of the fifty largest in the United States, was in those years.

The Engagement

The year 1934 was also special for me, particularly because of Hazel and Jerry Baron. As I mentioned before, Jerry was one of my golfing friends and we shared many activities together. He was a French Canadian from Maine and had many talents. Jerry served as a fighter pilot in World War I in Europe and was

It was at the Baron's that I met Harriet, a Minnesotan, in 1934. She had come to visit her Aunt Hazel Baron in Armuelles. After we became acquainted, Harriet and I made a three-day trip to Boquete with two other couples, the Prietos and the Chagnots — they were married, we were not. The trip was made by rail from Armuelles to Boquete and took about five hours. We stayed in a private boarding house in that small town located on the east slope of the Baru volcano at 4,000 feet, a nice, cool change from Armuelles. We rode horseback, picked some navel oranges, admired rainbow trout in a stream, and did some partying. Alberto Prieto, a doctor in the Armuelles hospital, was a young Panamanian with an M.D. from Tulane; his wife was a beautiful New Orleans girl. Alberto eventually set up his own practice in New Orleans and whenever we travelled through there, and when we lived there (1965-1969), he was our physician. Al Chagnot worked in the accounting department and his wife, Muriel, had been Mr. Blair's secretary before the quake.

After about six months of running the manager's office, I wanted to go back on the farms where I was eligible to be a district superintendent and where I would have a nice house so I could pop the question to Harriet. I told Mr. Blair this, and that I wanted a vacation for the honeymoon. He said he couldn't do without me in the office but that if I got married he would assign me a house in Armuelles, and give me another raise as soon as possible. He didn't see how I could leave on vacation but after awhile he arranged with KB to help him and thought that with a clerk and stenographer loaned from the accounting department his office would keep functioning for six weeks while I was away. I accepted all of these arrangements, Harriet accepted me, and thanks to Leon's holding back a couple of months' bar bills, I had enough money to begin the great experience of marriage.

On February 27, 1935, Harriet and I flew to Panama; I shopped for a ring, she shopped for a wedding dress (it was hot and I almost fainted waiting for her); and that afternoon we were married in the Union Protestant Church in the Canal Zone. Tom Jacome, the Company agent in Panama City, was the best man, and Esther Currey, a Canal Zone teacher friend of Hazel Baron, was the bridesmaid. Two days later we took a Company ship from Cristobal to New York, via Kingston, Jamaica, to visit my family in South Hanover and show Harriet some New England countryside. We returned to Armuelles on schedule and moved into a small single-family house which had been readied for us. So now we were together and on our own.

Married Life

We were comfortable in our little house — two bedrooms, a bathroom, a living-dining area and a kitchen, with a laundry room underneath. The Company furnished all first-class houses with a servant and a standard inventory of furnishings and supplies, depending on the size of the house, including an electric stove and an icebox. One of our first purchases was a Philco (vacuum tube) radio, shaped like a church front. All broadcasts were shortwave and the only ones available were the British Broadcasting Company from London and a U.S. Army station in the Panama Canal Zone, both with programs at certain hours only, and with questionable reception.

Each residence had a small chicken house and pen in the back yard. Our first "livestock" were some ducklings given to us by Lane Evans, a district superintendent on the farms. We bought native hens to supply us with eggs and some scraggly roosters to fatten up for *sancochos.* Later we would buy day-old chicks from the States and raise them, although it was a continuous and messy job to prevent, or cure them from, Newcastle disease. We kept chickens all the years we lived in the Tropics.

For a long time there was only one automobile in town, a beat-up old truck which served as a combination delivery, garbage pickup, fire engine, or any emergency transportation. The only roadway was from the wharf to the residential area, not quite a mile, and onto the water tank and dam. Everyone walked the short distances to work, to the club, the hospital, or the commissary although we could order and charge supplies by phone and delivery was made in the truck. Later Jerry Baron was assigned an old car for duty to and from the wharf and eventually even the division manager was furnished a second-hand car! Later another second-hand truck was brought in, fitted with a few fire extinguishers, a hose and small water tank to serve as a fire truck and driven to work by the master mechanic who was chief of the volunteer fire department. Before long, KB and a few other employees brought in personal cars, but there was nowhere to go except in the small town. There were no surfaced roads to the farms, or to David, in the years we lived there. Railroad motorcars (automobile chassis fitted with steel railroad car wheels) supplied transportation to the farms and on to David, but even from there the usual way out was by boat or airplane.

There was no completed highway to Panama City for several years and the Central American Highway didn't connect

Costa Rica with Panama until early in the 1960's. Almost all travel to Panama City from Armuelles was on a small, one-plane, airline operated by the owner Marcos Gelabert, a Panamanian pilot and his wife, Ida. (We flew to Panama with him on our wedding day.) Gelabert and I were good friends. He later was employed as a pilot by the Company in Costa Rica. Afterwards, he was killed in a crash while working with a small commercial line in Panama.

As in all Company divisions, where everybody knew everybody else, there was always a lot of social activity in the first-class part of town. There were "pecking" levels of course, but generally self-inflicted — either you wanted to party or you didn't. If you did, you were invited to a home, along with 20 or 30 others (if space permitted) every week or two for an evening of "cocktails" for a couple of hours followed by a lavish banquet-buffet setting of fancy foods with emphasis on the hostess' specialties. Then there would be highballs, record playing, dancing, and some efforts at harmonizing, often until the early hours. Then each couple "owed" the hosts and would reciprocate, inviting the whole gang to their house, so there was no end to it, and it was a lot of fun. Any Company guest visiting the division would always be included — often the reason for the party. The conversation, carried on by many small groups, was strictly Company business, gossip, lies, truths, and jokes. An outsider would not be able to understand most of the chatter. This life built a camaraderie that often lasted a lifetime.

Children of all ages enjoyed the same continuous swirl of birthday parties, generally out-of-doors on the lawn, with a plentiful supply of cake, ice cream, cookies, and candy and with small gifts for all.

All of this home entertainment was in addition to the many parties and dances held in the clubhouse. Saints' Days, States and Panamanian holidays, all were excuses for major celebrations by all club members. After work, if not swimming, playing tennis, or golf, bowling or sitting around the club, neighbors would visit one another for a couple of drinks and *bocas,* a little gossip and "radio listening." On Sunday evenings Mr. Blair and his butler would fill the half-gallon silver cocktail shaker (often more than once) with a Bacardi cocktail to accompany light snacks for a dozen or so guests and visitors. Since I was his *aide-de-camp,* we were always invited.

Artifacts, Butterflies and Fossils

With a servant and an occasional baby sitter in the house, Harriet and I often played golf on Saturday afternoons (during all my years in the Tropics everyone worked until 11 a.m. on Saturdays). On Sundays we would walk in the jungle or ride a motorcar to the farms to join Bob Cover in digging Indian graves. We found many beautiful painted clay pots and bowls but never any gold artifacts although many had been found in Chiriqui Province. The Chiriqui Indian antiquities anti-dated the Spanish conquest by several hundred years.

On our weekend walks into the forest we became interested in collecting butterflies. So that I wouldn't be seen with a butterfly net, I always asked Harriet to carry it through town until we got into the woods, and again when we returned back through town! In that lush tropical rain forest there were lots of beautiful species and we collected many, including some of the large blue morphos and owl butterflies. I would mount and pin them into glass-topped trays.

Since I had no idea what the scientific names were (there were no guide books in those days), I would airmail a series of each type, securely pinned in boxes, to Dr. Gertsch, Assistant Curator of Lepidoptora at the American Museum of Natural History in New York City. They were glad to get those Panamanian species for their collection, officially acknowledging them, and returning one specimen from each series to me with its scientific name. I also photographed the best one of each species in black and white and then made life-size enlargements which I would hand-color with special oil paints. Doing all that photo darkroom work in a blacked-out laundry room under the house in the Tropics, with no air conditioner or electric fan (which would kick up a lot of dust) was a hot job. All that photo work took hours and hours of my spare time. I had almost 150 life-size colored pictures of my collection, but they were all lost in the flood of 1954 in Honduras.

On those butterfly collecting trips I eventually learned that it was easier to get perfect specimens by collecting the catepillars and "hatching them out" at home, rather than frantically swiping at them with a net. However, I never found the larva of the beautiful and rare glossy metallic blue morphos until many years later at Lancetilla in Honduras. They were six inches long, banded red and brown, ½ inch thick, gregarious caterpillars which fed at night and spent the daylight hours huddled together in an almost solid clump in the highest treetops.

We had many thrilling experiences on those collecting trips: snakes, army ants, all kinds of birds (this was before my "birding" days), orchids, beetles, including the rare "goldbug" beetle, many small animals and some big ones. Once, while walking in the bottom of a deep ravine, a mountain lion came tumbling down from a couple of hundred feet above us. It had jumped at something and missed and when it hit the bottom, just a short distance in front of us, I don't know who was scared the most — the beast or us!

On one of our weekend walks in the rain forest, we came across an outcropping of fossil shells in the banks of the nearby *Rabo de Puerco* (pig's tail) Creek; well-named because it was really crooked. We collected a lot of them on several trips and kept them in a box at home. A friend, Robert A. Terry, a geologist exploring the area for oil companies, saw the collection one day and told us we had discovered a deposit of Pliocene and Pleistocene marine fossils. He contacted an associate, Dr. Axel A. Olsson, Associate Curator of Mollusks at the Academy of Natural Sciences of Philadelphia, an authority on the subject, and told him of our collection. Dr. Olsson was very excited about this and arrived in Puerto Armuelles a few weeks later. We took him to the "digs" where he collected for several days and then returned to the States with boxes of specimens, including most of ours. Later, in 1941, he and Dr. Pilsbury, published, in the Academy Proceedings Vol. VCIII, an account, with illustrations, of the collection which included *Chione Traftoni*, an extinct species new to science. "The type of this species (A.N.S.P. 13703) comes from the Pleistocene of *Quebrada Rabo de Puerco*, near Puerto Armuelles, Chiriqui Province, Western Panama, and is named for Mr. Mark Trafton, Jr., who was the first to collect in the Pleistocene beds at this locality. . . ." A copy of the Proceedings, with pictures of *Chione Traftoni*, is in my library.

Bill Sutherland, a Company headquarters lawyer in Boston, on temporary duty in Armuelles and Panama City, and his wife, Edna Mae, were close friends of ours. We had several common interests, among them pre-Colombian history of Central America and Indian artifacts.

San Blas

I was in Panama City for dental work and met Bill in the office there late one afternoon. We then went to the nearby world-famous Mamie Kelley's club for refreshments. Several hours later it was agreed that we would go to the San Blas Islands and take color movies of the Indians there. He was not a photographer and I had never taken pictures with a movie camera. We shopped until late that night — Bill spent about $800 for the best Kodak movie camera, film, light meter, and tripod. We also acquired a few trinkets for trading, several bottles of expensive brandy, a small supply of canned "food" (peanuts, guava paste, sardines, sausages, crackers) and two large towels.

In the morning we took the train from Balboa (Panama City) on the Pacific side of the Canal Zone to Cristobal (Colon) on the Atlantic side where we boarded a coastwise trading schooner for the islands. In our anxiety to get going on our expedition, we failed to advise the Panama City office or our families in Puerto Armuelles of our itinerary. I spent the daylight hours rocking along in the little boat, which had a crew of two, studying the directions for using the movie camera. I knew that if I didn't get some good pictures, my reputation as a photographer would be quickly smeared, and so would I! Bill had a violent temper!

His wife, a beautiful and stylish dark-haired Texan, had modeled for some of my first color pictures, taken on Kodachrome 4″ x 5″ sheet film with a large Graflex ground glass focusing camera. There was no 35mm color film in those days.

We arrived about noon the next day, went ashore in an Indian dugout toting our equipment and supplies. Our headquarters there, for our one afternoon, night, and morning visit, was in the open on the beach under a coconut palm.

The San Blas archipelago includes hundreds of idyllic small islands with white sand beaches, coconut palms, all in calm blue waters. Clans or small family groups of the Indians live on separate islands. They are remnants of the primitive indigenous San Blas Cuna Indian tribe. Fortunately the Spanish *conquistadores* missed this little strip of the Caribbean and, although the Indians are now considered Panamanian citizens, when we were there only a few of them spoke any Spanish and they dressed and lived in their own tribal ways. They were gentle, pleasant people and lived in thatch huts. Their food was mainly fish and shellfish, coconuts and some corn and vegetables which they grew in small plots on the nearby coast. The San Blas mountain range, unexplored primitive and wild country, rises abruptly from the coastal plain to over 4,000 feet. That was the setting for my first, and only, effort as a movie producer-cameraman.

When we arrived, the men were working their fields on the mainland but we were well-received by a group of about 30 women dressed in their bright, colorful skirts, blouses, and head-pieces. Many had gold rings in their noses. The *molas,* which the women make themselves by reverse applique, are large frontal and back pieces of their blouses. I took a particular liking to one which the girl took off and gave to me in exchange for two pieces of blue and yellow cloth (with which she could make another *mola*) and some colored postcards of Peruvian Indians with the shrunken heads of their enemies. Such were some of our trading trinkets.

A little later the "chief" arrived. He could speak a mixture of Spanish and English since he had spent several years of his youth as a seaman on a coastwise boat. Our friendship was cemented by a couple of drinks of brandy and he volunteered to assist us in producing a movie.

The scenario we had plotted was simple, but took several hours of rehearsing for the "star"(Bill) and the "extras" (six young girls in full native attire). I loaded and set up the camera and tripod on the beach with a view of sky, palms, sand, and blue water. The script called for Bill to be sitting alone on the white sand at waters edge, when out of the distance a dugout paddled by the girls comes into view. They land on the beach, jump out of the canoe, rush up to Bill, grab him and pull him (a willing victim) to the dugout. Then, with Bill in the canoe with them, they paddle off slowly into the sunset. End of Scene 1 — and several rolls of film.

Bill and I were offered shelter in a thatched hut but preferred to spend the night sleeping on the beach, with a towel as a mattress and a sky full of bright stars as a canopy. A little brandy helped to soften our "pads" and to warm the rather cool sea breeze. In the morning we joined the chief for a strange breakfast of pieces of deep-fried conch, some type of corn patties, and some "old" (fermented) coconut milk. Then, with his help,we were ready for Scene II.

On the far side of the island from the landing there was a round hut, completely enclosed with thatch, elevated on poles with a flimsy ladder stairway up about six feet to the only opening, a small door. We were told that young virgins, who had been betrothed by young "braves," were confined in this hut for several weeks before their marriage. No males were allowed to enter.

There were some colorful *molas,* with the usual abstract designs of birds, fish, and animals, attached to the outside, presumably to identify the girls inside. It was guarded by an elderly female. I started cranking away on the camera (no electric drive then) as Bill, a rather hefty person, approached the old woman, pushed her aside, and laboriously climbed up the ladder. I was afraid it would collapse, that he would fall and break some bones, and that we might become permanent residents of the islands, but it held together. After he entered, a lot of young girls started screaming and rushed out down the ladder. A few seconds later, Bill appeared in the little opening tugging a young maiden down the ladder. With considerable effort he carried her in his arms as they disappeared into a setting of palms, seagrapes, and blue sky. The closing scene was the chief sharpening his spear and shaking his head as he gazed into the distance. End of San Blas movie.

Our farewells were touching scenes as the schooner picked us up to leave in the afternoon. We had a smooth passage back to Colon but on the train back to Panama City the next morning, other passengers seemed to be staring at us. I guess we were slightly unkept. Needless to say that a lot of people were relieved when we reported in at the Panama City office. For four days no one knew where we were, families and friends were really upset, police had checked jails, hospitals, bars, hotels, passenger lists on planes and ships leaving the country, but no record of us! I was worried about Harriet and also afraid that I probably would be "fired" by the Company. But all actors, movie producers and cameramen have tough times and our lives returned to normal in a few days.

I only saw the films once and thought they were wonderful. They were rated "X" by Edna Mae, but Bill kept them for the rest of his life and I heard that after his death she had donated them to a museum on Indian culture in Houston, Texas, where she was living. I kept the *mola* for years and eventually gave it to my daughter, Sue, who cut off the sleeves and had it framed. So now she knows the true story connected with it!

I always wanted to return to the islands and early in 1942 Harriet and I, accompanied by John and Jessie Mase, made a special trip from Puerto Armuelles to Cristobal where we stayed in the Hotel Washington and made arrangements for a coastwise boat to take us to the islands. But by then the United States was involved in World War II in the Pacific. We appeared at the dock on schedule early one morning but the skipper informed us that the port was closed, with a submarine net across the entrance, and no traffic was allowed to enter or leave the canal. We soon found out why as we watched an almost endless convoy of all types of warships come into the harbor and enter the canal on their way through to the Pacific. We tried again the next morning, with the same results, so we never made it. No doubt the Indian maidens were disappointed.

On that trip, before going on to Cristobal, we took a boat to the beautiful resort island of Taboga in the Gulf of Panama where we spent a couple of days. Returning to Panama City we checked into the Hotel Central for the night. By chance, we met Reggie Hamer (Don Walter Turnbull's nephew, and "Picho's" dad) accompanied by Blondy Gronblad and his wife Bea, all from San Jose, Costa Rica. Reggie was General Manager of the Costa Rican operations for UFCO and was on his way to the Company division in Santa Marta, Colombia. I had met them previously, so this called for some partying, Company style, in my room where all seven of us started working on White Label highballs.

To prevent being a target for Jap or German bombs or shells, Panama City was completely blacked out, all window curtains drawn and the few taxis on the streets operating with no lights, and with a midnight curfew in effect, the city was almost ghostly quiet. After several hours of *cambiando impresionses* — discussing ideas — our voices were probably slightly raised since the desk called a number of times asking us to quiet down, which we would do for a few minutes while waiting for some more ice from the bar. Eventually there was a knock on the door and four Panamanian soldiers, all armed with rifles, said they had orders to detain us. Reggie, who spoke perfect Spanish, and was a polished diplomat, offered them drinks, which they accepted, and explained that he was on his way to Colombia on important government business, and showed them his plane ticket for early that morning. He also told them that the Mases and us were on a special mission to the San Blas Islands in the interests of Panama, and that he was leaving immediately to get his plane; that the rest of us were going to retire and there would be no more noise. After another round of drinks the soldiers, now our big pals, left and even ran down a taxi to take Reggie to the airport. After a couple of hours sleep, the Gronblads took the Company plane back to San Jose and with the Mases we caught the morning train to Cristobal. We didn't get to the San Blas

Islands, as I explained on the previous page, but the exciting several hours with Reggie and the Gronblads made the whole trip worthwhile.

Fishing and Boating

On Sundays and holidays Mase and I, with guests, would often go deep sea fishing in the Gulf of Chiriqui on the *Whiskey II.* A good catch was almost certain in the nearby rocky depths of Charco Azul and around Burica Point, if seas were calm. A specialty there was the large *papagallo* (roosterfish) weighing up to 70 pounds or more. They were not good eating but probably the fightingest fish in the ocean when hooked trolling with rod and reel. We also sighted, but never fished for, sailfish and giant blue marlin around Burica Point. I thought the best eating fish were the beautiful dolphins (the fish, not the mammals) but there was always plenty of large kingfish, Spanish mackerel, barracuda, tuna, and red snapper, some of them weighing up to 30 or 40 pounds. Sharks were a problem, often biting off all but the head, and sometimes the whole fish, rig, leader, and some line before the catch could be boated. Mase and I once came upon a beautiful large spotted tiger shark which swam alongside our boat for half an hour. We estimated it must have been forty feet long.

In season, there were huge schools of large jacks, the females heavy with roe — up to two pounds. We would use hand lines with spoons to catch them right beside the boat, discard the males and slice the females open to get the roe and throw the bodies overboard. Before long the bloody mess in the water attracted lots of sharks and a feeding frenzy resulted which made it dangerous to even try to pull a jack into the boat — a shark could easily take the fish, and an arm! In an hour or so we could almost fill a washtub with roe and, back in Armuelles, the messhall chef would fry it up for a special afternoon snack for all comers — after John and I gave liberal portions to friends and took home all we could handle in our freezers. It was delicious, a treat for gourmets.

On one occasion, when the sea was dead clam, almost glassy, there were thousands of sea snakes about two feet long squirming around on the surface as far as we could see in all directions. It was a beautiful sight with their black bodies with lines or spots of yellow and their reflections in the silvery-blue calm water. To observe them close up, we launched the skiff and paddled through some, dipping an oar under them and letting them slide down into the skiff. They are deadly poisonous and there is no antidote for their bite. We kept them alive in a bucket until we returned home and preserved them in large jars of alcohol.

Mase had a copy of an old *(authentic?)* Spanish pirate's log indicating he had buried some treasure on one of the small *Islas Secas* when being pursued by some English ships. We had a Goldak meter (metal detector), a large contraption with a transmitter-detector on one end and a receiver on the other end, which a person could carry on a sling. It was a far cry from the present-day detectors seen on beaches everywhere. We made several two-or-three-day boat trips to the islands, a ten-hour run from Armuelles, and laboriously criss-crossed them each time with the meter, chopping our way through bush and covering the beach at low tide. It was hot work and in the evening we would dive for some large conchs and stew the meat with potatoes and onions for supper, while discussing the next day's plans over a few drinks while watching the sun set.

On our third voyage we hit pay dirt with repeated strong signals over a spot on the sandy beach. So we started digging, alternating fifteen-minute shifts which was all we could take in the hot sun. On the second day we uncovered what was surely one of the claws of a large forged anchor from a very old ship and the top of an iron barrel type chest. The trouble was that down six feet in the sand on the beach, the hole at that depth kept caving in and filling with water, even at low tide. So, after getting a bearing on the spot, and with the White Label supply running low, we reluctantly sailed back to Armuelles to plan how we could some day return with a small dredge. World War II ended those dreams and John passed away several years later. However, for a satisfactory renumeration I can mark an "X" on a map of the Islas Secas (if you happen to have one) showing the approximate spot on the beach, on one of the several islands there, where the treasure can be found. Otherwise the secret of the Trafton-Mase treasure will die with me!

Another "fishing" story is an account of a local vacation we took with Sue, Mark III, and the Mases at Bambito, located at an elevation of 4,500 feet on the slopes of Baru, the 11,400-foot-high Chiriqui volcano. Two German-speaking expatriates, "Slim" Cowart and Kurt Hammerling, an Austrian, owned a large property there along the headwaters of the Chiriqui Viejo

River, a narrow, rapid-running cold mountain stream with water cascading over large boulders. Close to the river they had built three small cabins, in addition to their own, which they would rent to selected guests, mostly from the Canal Zone. Except for the barren path of an old lava flow from the peak down to the middle valley, about a mile downstream from our cabin, the whole area was blanketed by a typical montane rain forest with a canopy of magnificent straight tall oak trees, many over 120 feet high.

A former United States Ambassador to Panama had arranged for stocking the mountain stream with rainbow trout from the States. They adapted perfectly and soon provided some thrilling dry fly fishing. I had acquired a fly rod, reel, line, creel, and a collection of dry flies — "Royal Coachman," "Silver Doctor," "Light Cahill," and others. During our first few days there, Slim taught me the fine art of dry fly fishing and I became quite proficient although some of my flies are probably still embedded in tree branches along the stream. Mase and I would cast in front of our cabin early in the morning when a "hatch" of midges were rising out of the water and in about half an hour we would have five or six good-sized rainbows for breakfast!

One day I was fishing alone, working slowly upstream and, around a bend, a beautiful young girl was standing on a rock and casting. I couldn't believe my eyes, I was transfixed, but I hurriedly took the land route back to my cabin for my camera. She and her sister were vacationing with their parents further up the mountain. We introduced ourselves. Her name was Manonguita from David. Thereafter we met several times, fishing.

A few days later, the fishing editor of *Outdoor Life,* Bob Fackelman, arrived on the scene to do a story "Trout in the Tropics" for his magazine. Later one afternoon, Slim and I accompanied him just as the evening hatch was rising. In about an hour before dark Slim had fifteen rainbows, I had twelve, and the "expert" had none! At the cabin I took flash pictures of the catch. He didn't do much better the next morning — he had three while Slim and I had about twenty-five. His article mentioned my name several times and appeared in the December 1941 issue of *Outdoor Life* but he couldn't publish the pictures because the "catch" was way over what the daily limit was in the States.

The kids loved this life in the wild, except for being scrubbed down in the cold water. Mase and I roamed the forest collecting orchids, observing a nesting pair of quetzals close to our cabin (a lifetime thrill for me) and vainly hunting for *tigres* and *javelina* pigs. We did shoot some of the large band-tailed pigeons for the larder. It was here that I collected the rare "goldbug," a large beetle with an overall brillant gold coloring. It was a wonderful vacation but it could not be repeated. A passable gravel road was soon finished through the area and the majestic forest was cut down to plant pastures for grazing cattle. Heavy rains soon washed the soil away leaving only a barren, sterile, rocky terrain. The polluted stream could no longer support the rainbow trout and they disappeared, victims of man's thoughtless destruction of the environment.

Sigatoka Invades Central America

In 1935, the dreaded banana leaf disease, Sigatoka, appeared in Honduras for the first time. Sigatoka had wiped out plantations in the Far East and now threatened to do the same to our bananas in the Western Hemisphere. There was no known cure but Sam Zemurray's confidence in Dr. Dunlap as Tropical Research Director paid off. He had no time for controlled experimentation but Doc reasoned that a copper sulfate, lime and water mixture, known as Bordeaux, might control the fungus. Preliminary tests showed promising results and Sam Zemurray gave the order to start spraying to save the banana industry.

The big question was how to mix millions of gallons of Bordeaux and spray 150,000 acres about every two weeks. Company engineers and research people developed a plan and soon a 5,000-acre experimental project was installed at a cost of half a million dollars. Because of the urgency, Zemurray didn't flinch at this expense. To cover each 1,000-acre area, a mixing and pumping plant was installed with two 2,000-gallon wooden mixing vats and a huge high-pressure diesel pump. A maze of pipes was laid in the banana plantings to attach hoses for spraying the mixture onto banana leaves. This plan worked and was installed throughout all Company divisions in Central America, thus saving United Fruit from destruction by this new disease.

By early 1938, Sigatoka was detected in Panama and Doc Dunlap, my old pal from Esperanza days, requested that I make a trip to the Research Department in Honduras to familiarize myself with the disease and its control measures. Mr. Blair in Armuelles and Walter Turnbull, the General Manager of Tropical Divisions, agreed. Besides making intensive laboratory and field studies of Sigatoka in Honduras, I observed the problem in the Guatemala divisions at Tiquisate and Bananera.

I returned to Panama, and on March 1, 1939, was appointed to the new position of Superintendent of Disease and Insect Control. This included a raise to $3,000 a year, a motorcar and motorboy to go back and forth from Puerto Armuelles to the farms every day. Two qualified assistants from Honduras, Ira Hubbard and Rudy Lunemann, a young German, were assigned to me. I hired Crandall "Chips" Spence from California, a brother to Ed who was a farm overseer in our division.

As we installed spray systems throughout the division farms, I made hundreds of experiments and observations on application and control techniques. A method was designed to achieve better control by spraying with an overhead system rather than using a hose spray underneath the leaf canopy.

Having been interested in butterflies and moths, I began to photograph caterpillar pests and their types of leaf damage, and then hatch the pupae to photograph the adult forms. Each picture, which I would enlarge to life size and color by hand, was then mounted and included with a complete text with scientific names in a manual titled "Caterpillar Pests of the Banana" which is still used as a reference. Because of my knowledge of insects and my success in developing Sigatoka control methods, I was later promoted to Plant Disease and Insect Control Superintendent for all divisions of the United Fruit Company.

New Changes

Armuelles was now undergoing changes. Gardner Myrick (the big boss), a Zemurray man from Puerto Cortes in Honduras, was made manager and Mr. Blair was transferred to Panama City as Company agent and later retired there. K. B. Block was transferred to the Golfito Division in Costa Rica as Superintendent of Agriculture, but soon became manager. Andy Holcombe was made superintendent of farms in Armuelles. I weathered those losses of close friends but got along well with the new chiefs, mainly because neither of them knew anything about Sigatoka and depended on me to keep the disease under control (and their jobs safe!). Myrick was a rough and tough *hombre*, formerly the Port Superintendent in Puerto Cortes, and was under orders to bring the division in line with Zemurray's policies so there were quite a few casualties among personnel. I not only survived but, before long, Myrick and I were good pals.

Since spraying for Sigatoka control ended early in the afternoon due to winds, after lunch at the district messhall I would generally check experimental plots I had laid out and then return to Port, not to the office, but to the La Playa commissary in the labor camp section of town. The black commissarian there, Mr. McBean, made a wonderful Scotch highball and the Trainmaster, our next door neighbor, Slim Banack, and I would sip a couple before the four o'clock whistle blew, preparing us to play golf. Myrick soon wised up to this ploy and from then on was always there, even before I would arrive. He would ask me how I could run my job from the commissary, and I would counter with how could he manage the division from there. My answer was, "It was a waste of Company money to spray when the wind was blowing and that I was out working early in the morning while he was still in bed." His answer was that, "He had just fired a couple of people so he didn't have much to do!" Those statements, always in fun, became our greeting whenever we met. Slim Banack didn't have to worry about his job either, so long as the trains of loaded banana cars always got to the wharf on schedule.

Some Interesting People

Panama has been called "The Crossroads of the World," and with good reason since people from all over the world go through it. Many stop or remain there. I met a few since some of them were going north, or south, on Company banana ships; others drifted through the Miramar bar in Armuelles looking for hand-outs.

Henri Charriere, a Frenchman, convicted of murder and sentenced to life on Devil's Island, the French penal colony in French Guiana, was one of the strange people I met up with. His nickname was "Papillon" from a large butterfly tatooed on his chest. Escape from the island was always considered impossible, but he did! He arrived in Armuelles broke, and needed funds to take a Company ship to California. I spent parts of several nights in the Miramar Club listening to his hair-raising accounts of cruelty to prisoners by the guards on the island, the end of some of his jailmates in a bloody swirling shark feeding frenzy in the ocean just off the beach, and of his escape. He accepted cash donations with the promise to send a copy of the book he was going to write to all donors. He got a few Balboas from me; he did write a book, *Papillon,* which became quite popular; but he didn't send me a copy! A movie of the story was very successful.

Albert Einstein, and his wife Eloise, were on a Company ship from Panama to California after moving to the States from Germany. During their stopover, Mr. Blair asked me to meet them in Esperanza and show them the banana farms. I took advantage of the occasion to go there the night before to visit with friends, among them Duffy Reagan, brother of the accountant. (They came from Taunton, Mass. where I once played with an orchestra in the Roseland ballroom). We celebrated most of the night and the next morning it was pouring rain but the Einsteins arrived there by motorcar, with the storm curtains buttoned down on all sides. I had a *goma*, hangover, and so did Duffy. He rushed out to the motorcar in the rain, raised the curtain, stuck out his hand and said, "Hi, Al, Reagan's the name" and then returned to the bachelor quarters. I got in the motorcar on the seat between Einstein and his wife to ride back to Armuelles with them since we couldn't look at bananas in the rain. I felt very ill but managed to point out things of interest along the track. His English wasn't much better than my German; he knew nothing about bananas and I knew nothing of the theory of relativity except E=mc2, but we smiled and nodded all the way back to Port. What an impressive person, that head of frizzly grey hair; those wide-open eyes which had looked into and fathomed secrets of the universe; this appreciative, unassuming manner — a true scientist.

A Belgian adventurer also created some talk. Much of the area around the heavily forested Chiriqui Volcano, over 11,000 feet high, was inaccessible but there had always been stories of old Spanish silver mines and caves where they had stored gold Indian antiquities. They would accumulate a large cache in a remote cave in the mountain awaiting the arrival of a ship off the coast, a year or so later, to transport the treasures to Spain. With that history, all kinds of fortune hunters were attracted to Chiriqui Province and Armuelles was the supply and take-off point for many of them. Many grizzled old men, typical prospectors, would hang around the Miramar cantina in Armuelles, swapping wild tales of the riches they never found in exchange for a few drinks and a few Balboas to start them on their way again. I contributed to many of those "grubstakes."

A Belgian said he had discovered an old Spanish silver mine, with piles of silver ingots stored back in a long tunnel. This tale quickly got to Panama City where it became newspaper headlines. Government officials soon sent a large force of the *Guardia*

Nacional with trucks, tents and all kinds of equipment to the nearby city of David. With the Belgian in their custody they set out on foot with pack mules loaded with supplies and the Belgian guiding them to the now highly publicized rich cache of silver. About two weeks later, they returned empty-handed and without the Belgian. He had eventually led them deep into a dark tunnel where he shot himself — so they said. There were no silver bars and no trace of a silver lode. The Spaniards must have cleaned it out! The Belgian, in trying to rustle a few Balboas, had made the mistake of spinning a yarn so intriguing that it cost him his life!

Jimmy Angel was a well-known "bush pilot" whom I had met on several occasions in Armuelles. He made a living flying adventurers in his small single-engine plane around Central and South America. On one trip from Panama to Colombia and Venezuela, his passenger was a fellow who claimed to have found a rich gold mine, and had plenty of nuggets to prove it. He paid Jimmy well and directed him to a remote area high in the mountains of Venezuela where he pointed out the general area as they flew over it before returning to Panama. Unfortunately the man soon died of malaria but Jimmy was so convinced that he moved to Venezuela and made many flights by himself hoping to locate the area again so he could plan an overland trip through the jungle to gather up some nuggets. He never found it but in flying over those remote and unexplored Venezuelan mountains he discovered the highest waterfall on this continent, now officially named Angel Falls.

Hollywood movie stars often made the "banana boat" trip from Los Angeles to Puerto Armuelles and return. One of them was Pat O'Brien who, with his wife, visited with Pop and Mom Diebold (Pop was the master mechanic in Armuelles) whom he met on a previous trip. Pat acted in many popular movies and co-starred with artists like Bing Crosby, Jimmy Cagney, and Ronald Reagan. The O'Briens visited Armuelles several times, and, of course, were the honored guests at all the local parties during their visits.

In 1941, Harriet and I were on a weekend visit to the mountain town of Boquete and, walking down the dusty, rocky road to our *pension* in the evening, encountered a couple, obviously tourists. As we passed, the man said, "Dr. Trafton, I presume? I've been searching for you." He was Victor Wolfgang von Hagen, a German who had written several popular books on Latin America, including *Quetzal Quest*, of which I had a copy, and *The*

Green World of the Naturalists. His wife, Cristina, a trim, blond California girl, also authored several children's books, as well as proofread her husband's books — he was pretty terrible when it came to spelling. They were gathering material for a story on Panama and had been referred to me as a "local naturalist" when they arrived in Armuelles. They were staying in the Company guest house there but just about did us in by arriving at our house to visit every afternoon about cocktail time. They always stayed for supper, and for several hours afterward, bleeding us for material for his book. He was an impossible bully and made a hasty departure from Panama without his wife when the FBI started picking up Germans when the States declared war on Germany. I loved Cristina, a truly gifted and gentle girl, who stayed on our free meal list for some time after he left. Meanwhile, she wired home for money for steamship fare back to California. She divorced von Hagen soon afterward.

The War

Although World War II had been going on in Europe since September 1939, it had not greatly affected the United Fruit Company or our way of life. However, there was a sudden change on December 7, 1941, when I heard on the radio that Japan had bombed Pearl Harbor and wiped out most of the Pacific fleet. The States declared war on Japan immediately, and on Germany and Italy on December 11. Rudy Lunemann, my German assistant, was in our house when this news came through. He said, "That's it, I hate to see our friendship end but I'm resigning and returning to fight for my country." And he did. I admired his patriotism. Several years after the war, I learned that he had commanded a German panzer (tank) division; that he had survived; and returned to live in his native city of Essen, which had been heavily bombed by the Allies.

Of course, the war news shocked all Americans but the few of us, and the Panamanians as well, in Puerto Armuelles had greater fears. The Jap fleet might continue sailing east and hit the Canal Zone, which would have been a sitting duck. Alternatively, they could land unopposed in Armuelles, wipe it out, and use the airfield there and in David for a siege of the canal by land and sea. On the day of Pearl Harbor, a squadron of U.S. fighter planes from the Canal Zone were parked in Armuelles while the pilots were playing golf with us. The planes were not armed. A few planes in the Canal Zone had the single machine guns mounted, but had no ammunition!

Like everyone else in town, I was afraid and, with two small children, Mark and Sue, very worried. We actually made up a "survival kit" of food and clothes which we kept in the back entry in case we had to suddenly take off into the jungle. Shipping ceased on the west coast fearing Jap submarines. All of the young Texan banana farm overseers, former ROTC students, had to return to the States and join the armed services. I eventually received a draft card, but with a temporary deferment due to age and family.

With the States now committed to a large-scale war in Europe, and in the Pacific on the other side of the world, the Company's largest and fastest ships were requisitioned to transport refrigerated cargo to bases in England and the Pacific. Smaller, slower banana ships, which could not out-run submarines, were kept in service for a while on a greatly reduced scale, so the future of the banana industry didn't look too good. Also, I had a strong conviction that I should be doing something to contribute to the war effort. In the Canal Zone the U.S. had already started work on a second canal and set of locks and this project now became especially urgent. There was a desperate need for skilled middle-aged supervisors in all categories because the younger men were being drafted into the armed forces. I had no connections in the Canal Zone, and would have no money except severance pay, but in March 1942, I wrote a letter of resignation from the Company and sent another to the Panama Canal employment office. I had hoped to get work and settle down in the new Diablo Heights section on the Pacific side of the Zone.

Myrick, the manager in Armuelles, radioed Don Walter Turnbull, the General Manager in La Lima, for authority to give me severance pay. In less than a week, Don Walter arrived in Armuelles by Company plane and asked me to withdraw my resignation and accept a transfer to La Lima, Honduras, for assignment in the Research Department there where Doc Dunlap wanted me as his assistant. Don Walter told me that Zemurray intended to keep the Company in business, one way or another, and maintain key divisions for the duration. I would be placed in charge of disease and insect control on all 175,000 acres of Company bananas in Central and South America, and in the Caribbean Islands.

I explained my wish to help, one way or another, in defense

work. He told me I could make valuable contributions in the agricultural field through the Research Department since Zemurray had offered the United States government all-out assistance in growing strategic crops essential to the war effort such as rubber, abaca, and some essential oils no longer obtainable from the Far East. I accepted, and was also greatly relieved, being able to stay on the payroll and at the same time "do my bit." That sounded much better than trying to maintain my wife and children while job hunting in the Canal Zone!

The transfer was arranged. We packed up and left Armuelles on April 16 via motorcar to David where we piled into the rear of a station wagon with a few suitcases for the overland trip to Panama. Most of our belongings had been sent by ship to Panama and on to Cristobal where we would connect with a Company vessel to Honduras.

The ride to Panama City was an ordeal, traveling over the proposed Pan-American highway — rocky, rough, and dusty. Fortunately, it was the dry season and the rivers were low enough to drive through the many which had not been bridged. From David, we drove all afternoon and all night before arriving at La Venta (Rio Hato) where we stopped at a small motel on the beach for a two-day rest. On the way, we had stopped at night to eat in a dirty roadside joint where a lot of fleas latched onto Harriet making her very uncomfortable for the rest of the ride. From La Venta to Panama City the road was somewhat better, and we made that stretch in fourteen hours. We spent one day in Panama City and two days in the Hotel Washington in Colon waiting for our ship, the *Santa Maria*, which would take us to Honduras. On our last day in Panama City, we had stopped in an ice cream parlor for snacks. The kids and I had ice cream and soon became ill with stomach problems. Harriet had no ice cream and no problems.

Security and customs were very strict in the Canal Zone but luckily we didn't have to open all our crates and packages since the customs inspector assigned to us in Cristobal turned out to be an old school classmate of Harriet's.

So for me it was *adios* to Panama after fourteen eventful and very pleasant years.

With three of us sick, our first night at sea was not a pleasant one. To make matters worse, the ship was on a continuous zig-zag course, trying to make it difficult for German submarines to torpedo us. We arrived in Puerto Limon, Costa Rica, the next morning and spent the day in port. Several weeks later, on July 2, 1942, another Company ship, the *San Pablo,* was torpedoed while tied up to the same Company wharf in Limon.

After dodging submarines all the way, two days later we arrived safely at the Company port of Tela, Honduras, where we were met by Dr. E. Otto Reif, Chief Chemist of the Research Department. We were put up in an apartment there but Mark III was still ill and spent the night in the hospital. The next morning I took the Company passenger train to La Lima (5 hours) where I stayed with Doc Dunlap and his wife, Lydia, for a couple of days getting squared away. Then I returned to Tela to take Harriet and the kids back to La Lima where, it turned out, we would live for twenty years and add Wilbur, Janet and Margaret to the Trafton family.

A Nurse in Port Limon

By

Laura B. Smith Kreiton

While working at New York Hospital, I took care of Dr. Conner who was head of the Medical Department for United Fruit Company. After his discharge from the hospital, he phoned me out of the clear blue sky to ask if I would like to work for the U.F.Co. in Costa Rica. Things happened fast. I made an appointment to see him the next day and soon signed a contract to stay one year as a nurse in Port Limon. A week later in that month of May, 1936, I was on a banana boat, the S.S. *Veraguas*, headed for the Tropics. Needless to say, the week before sailing, I was busy beyond belief. It wasn't until I was on the boat with time to think that I began to realize just what I had done. No doubt, I would honor my contract and stick it out.

Upon my arrival in Port Limon, I found the Fruit Company people all very friendly and helpful in getting me through customs. Afterwards, they took me to the medical compound where Dr. Edward I. Salisbury was head of the hospital. He later followed Dr. Conner as head of the UFC Medical Department in New York City.

One of the first things that impressed and astounded me the morning after my arrival was to see a man riding on a donkey with milk cans strapped to either side of the animal and delivering milk to the houses around "The Point" where we lived. The milk would be dipped out of these cans and poured into containers brought out to him by the cooks or maids from each household. Having been no further than New York City from my hometown in western New York state, it was surprising to learn that water and milk had to be boiled before we drank or used it. Also, we could not eat any fresh vegetables raw — they must all be cooked because of amoebic dysentery.

My new home was in a house with Kate Donegan, Head Nurse, who had worked with United Fruit for some time, and three other nurses. I was impressed by the fact that we had a

---

Laura B. Smith Kreiton was born in 1911 in Castile, New York, and became a nurse in Rochester. While working in the Costa Rica's Port Limon hospital in 1936, she met her husband, Benjamin Kreiton, who worked in the Accounting Department. They returned to the USA in 1938. Mrs. Kreiton retired to Port Charlotte, Florida, the original site of the *Quince* reunions of United Fruit Company people.

cook, a maid, and a houseboy to wait on the five of us. There were three brilliant Costa Rican doctors on the staff. Two of them were single and ate their meals with us nurses in our home.

The hospital had one hundred beds, an operating room, and an emergency room, which also served as an outpatient clinic. Connecting the O.R. was a central supply room where the O.R. nurse wrapped dry goods, fixed solutions and did all of the sterilizing for the O.R. and E.R. One ward had about fifteen beds for chronic patients who were no longer able to work. It was in this ward that I saw my first and last case of elephantiasis. All inpatients or outpatients treated in the hospital had been or were employees of the UFC or were members of the employee's family.

Our work in the hospital was interesting, consisting mostly of snake bite wounds, malaria, and accidents. Malaria-carrying mosquitoes were out in the bush and those people living out there were most apt to get malaria. At two different times, men came into the hospital with an ear in their hand that had been sliced off close to their head by a machete. The ear was sewed back on by one of our competent doctors and was once more a "live" ear. It was the first time I had seen that done or even heard of it being done. When the victim was asked who cut his ear off, he would reply, "a friend." Some friend!

We were not supposed to accept pregnant women of nonemployees who came to deliver their babies unless there were complications. I was on night duty when a woman in this category came in ready to deliver. As a nurse, I had to call the doctor to examine her before she could be sent home. The doctor came immediately but by the time he arrived, her delivery was imminent. She was delivered there and stayed overnight. The doctor asked her why she came in, particularly since it was her fourth baby. Her reply was that she did not want to disturb her husband. She hadn't for she had come alone by taxi.

In the beautiful UFC compound at "The Point," bordered by the sea wall, were a good number of houses for the married people and their families. Most of the single men and women lived on the second floor of the Community House. The first floor had a kitchen, bar, and tables where those employees who lived there were served their meals. We had a large recreation room with a piano, ping-pong table, and a pool table which I particularly enjoyed along with the others.

At least once a month, there would be a dance in the Community House, preceded by dinner being served to different groups in different homes about the compound. For these events, the ladies always wore long gowns and the men dressed in white suits, white shirts, and a necktie. It was informal and yet we were always dressed up. Some of the upper class Costa Ricans, especially the girls, would attend but the girls were always well chaperoned by mama. One nice thing about these parties was that the nurse on night duty was allowed to go to the dinner and dance, but was "on call." However, she had to go on duty after the party was over. Never have I worked anywhere else where I was allowed that privilege and luxury!

Of the five nurses in our house, two supervised the wards with their patients, one covered the emergency and operating rooms and one nurse was on night duty. We all rotated these duties except Kate, the Head Nurse. We five R.N.'s supervised the native personnel who did the bedside nursing.

The operating room was on the first floor of the two-story building and there were screened, open windows in almost a semi-circle about the room. During an operation, about one-third of the lower half was closed by shutters, while the upper two-thirds was open to the outside with just the screen to keep out insects. To me, this seemed unsterile and I expected wound infections, but we had none. There must have been something about that good, clean, tropical air along the sea that prevented infection.

Downtown Limon was different. We were not to eat there for fear of amoebic dysentery but we did on rare occasions. Before, I had only seen picturs of a community laundry with big wash tubs and the women leaning over wash boards scrubbing their laundry by hand. It was all reality there.

There were scarcely any roads outside of Limon. We went to San Jose by railway through rough, mountainous terrain or flew over the same terrain by plane.

At the end of my first year, I took a one-month vacation and was most eager to go back to Costa Rica for a second year. I met many friends there with whom I still keep in contact. It was there that I met my husband, Benjamin Kreiton, who worked in the Accounting Department. We both left Limon in 1938, having been there for two years each.

Many of these memories are relived each year at the United Fruit Company reunion, known as *The Quince,* which is held here in Port Charlotte, Florida, where I retired.

Tico in the Tropics

By

Oscar Chavarria Aguilar

I worked for the Compañia Bananera de Costa Rica on two separate occasions in the 1930's. The first time was during a summer vacation — our *verano* was December, January and February — when I was assigned to the Coto-Colorado Valley in southern Costa Rica on the Pacific side where the Golfito Division was just beginning. It was late 1937 and I was just barely a boy of fifteen years old.

The second time I worked for the Company was a year or so later as a stock control clerk in the Materials and Supply Department at Puerto Cortes, the base camp on the Rio Terraba which was the launching point for the development of the Palmar area. I worked under Mr. Elmer LeBlanc, a likeable, somewhat phlegmatic Nova Scotian with a beaky nose and a jutting jaw that always reminded me of Punch of the Punch and Judy show.

After a couple of months at Puerto Cortes, a number of us were transferred up river to Palmar Sur which would later become the center of operations. Upon arrival, we discovered that we were the first occupants in this virgin territory. Building and development was just getting started.

In the beginning, there was nothing to do in Palmar Sur except work. However, there was Palmar Norte across the Rio Terraba. Palmar Norte was mostly a small village of Boruca Indians. Inevitably, as banana people came into the new region, a couple of cantinas appeared. One cantina even boasted of what they called a dance hall. Since there were almost no women to dance with, the alternative was to drink. When we wanted to cross the river to Palmar Norte for a few beers on a Saturday evening, we hollered to one of the Indians lolling about in his dugout waiting for just such a summons. The Indians would

---

Oscar Chavarria Aguilar was born in 1922 in Costa Rica and worked as a youngster in the Golfito Division in the 1930's when the area was a wilderness. After the outbreak of World War II, he joined the U.S. Army Air Corps. With the help of the GI Bill, he studied at Berkeley and then got a doctorate in linguistics in 1952 at the University of Pennsylvania. Dr. Chavarria was on the university staff in Michigan, Rochester and City College of New York where he was Dean of Liberal Arts and Science. In 1977, he retired to Costa Rica and lives in San Isidro de Heredia.

ferry us across the river for a rather stiff fee. It was decidedly a seller's market: no ferry, no beer. Life was, on the whole, placid and uneventful, apart from an occasional drowning or a knife fight.

It was my first job with the Company that I recall most fondly, a job I got through my mother's cousin, Chale Quiros, who worked for the Company then. For a couple of months, I was rodman with a survey crew that was mapping the area for the first rail line which was to be constructed between the new port of Golfito and Palmar some fifty miles away at the other end. Conditions were indeed primitive as we were in dense jungle most of the time, miles from nowhere. Also, the work was hard. But for a fifteen-year-old, it was a lark! (I didn't feel so larky when I came down with malaria on my way back to school though.)

Getting from our homes in the highlands of Costa Rica to the jungles of the southern lowlands was quite an adventure in itself. Our group of new employees went by train to the Pacific port of Puntarenas and then boarded a rickety old tub, called *Centroamerica.* Somewhere off Punta Judas, the boat hit something and sprang a leak. The passengers had to take turns with the crew manning an asthmatic old pump and bailing for dear life with anything that would hold water. We put into Parrita for repairs and eventually made it to the Company's temporary headquarters at Puerto Jimenez on the Peninsula de Osa without further mishap. At Puerto Jimenez, we boarded a smaller launch for the trip acorss the Golfo Dulce and up the Rio Coto. When the river was no longer navigable, we threw our meager belongings onto our backs and continued on foot to our destination. Actually, my belongings were not so meager because my mother even packed sheets for me. We hiked for two or three days through virgin jungle before we reached our base, a rough clearing in the forest called La Niagara.

Niagara was not well-named because there were no waterfalls to speak of. The camp consisted of three principal structures that were thatched with palm fronds. These ranchos were without outhouses, I might add. The largest structure was the nicest and this was where the engineers lived. This rancho had walls made of palm trunk slats and was even equipped with a kerosene refrigerator and a mess facility for cooking and eating. The three engineers were the Odio Cooper brothers — Carlos Manuel, Orlando and Enrique — and they may have been Cubans. They shared their sumptuous quarters with a mutt named Hatuey.

The second palm thatch structure was without walls and was for us middle class rodmen and chainmen. There were about ten of us under this roof. The third palm hut, also without a wall, was for the peones. The middle class and the peones shared a kitchen together but we ate separately. This was mainly because we ate on our bunks inside the mosquito netting to get away from the tormenting insects. Our bunks were a sort of stretcher made by poking two poles through a couple of gunny sacks which was suspended about three feet above the ground. This rig was hung on four forked sticks driven into the soil. Our floor, of course, was the pure earth. Surrounding this little camp was virgin jungle.

After I'd been at La Niagara a month or so, a serious engineering problem arose. The original route for the rail line would have gone over a considerable gorge which would have necessitated a long and costly bridge, so the Company decided to seek an alternate route. They brought in an engineer, Mister Edward Crawford, from Honduras. He was flown from the little air strip at Puerto Jimenez to make an aerial inspection. We could see the plane swooping and circling as we sloshed around on the ground below. After studying the area by air for a couple of days, the flights ceased, and we waited. A few days later, the engineer from Honduras showed up at La Niagara and started an on-the-ground survey.

Mr. Crawford was a slight, rather jumpy man with a shock of quite unruly, red hair. He chain-smoked Camels. We hardly knew his first name because we all properly addressed him as Mister Crawford. Mr. Crawford asked for and got the services of a guide and an interpreter since his Spanish was marginal, to put it politely.

The guide was a *baqueano,* the best in the camp, whose name was Matus, a Nicaraguan. It had not been very long since the original "Tacho" Somoza had seized power in Nicaragua and many of the losers who had fought under Sandino escaped from Nicaragua. Hundreds of Nicaraguans were brought to the developing banana divisions to open up the wild country. It seems that many of them claimed to be colonels under Sandino; at least, I never met one who admitted to a lesser rank! In any case, Matus was a good guide. Had he lost his way back to camp, though, we would have picked up the trail just by following Mr. Crawford's cigarette butts.

I was named as the interpreter, a job I welcomed since it got me away from the third Odio brother for whom I was rodman. He and I had a personality conflict and he was always onto my tail for one thing or another. There was another reason I was glad to be interpreter. My father and grandfather had both been civil engineers and it was taken for granted in the family that I would follow their footsteps. I never knew why since I could never add two and two. Instead, I became a linguist. I must have mentioned those great expectations to Mr. Crawford for he was always willing to explain things to me and to show me how the instruments worked — a thoroughly nice man. He even let me read the altimeter on occasion; I felt like someone special.

Every morning, literally at the crack of dawn, Mr. Crawford would tell Matus, through me, what he was looking for and where he wanted to go. Then the three of us would set out — Matus in the lead, Mr. Crawford in the middle, and me bringing up the rear. My belt was festooned with the instruments which Mr. Crawford ceremoniously handed me each morning and which I would hand back to him during the day upon request. He usually said, "Oscar, hand me the *chingadera,*" and I would reply, "Which *chingadera,* Mr. Crawford?" It was usually the altimeter.

On the last day of Mr. Crawford's reconnaissance, we were coming down a long, steep hill. Matus suggested that we could make better time if we took advantage of a thick vine that would swing us down and across an embankment. The guide went first, going down and then making a large arc that landed him way off to our left. Mr. Crawford then gleefully followed suit, yelling with what I took to be some primitive Irish war cry, never losing the perennial Camel between his lips. Then it came my turn and disaster struck. As I started my swing downward, the vine partially gave way some twelve feet more than the other two men had started with. As I came whooshing by, I clipped Mr. Crawford in the back of his knees and knocked him to the ground. "*Dios mio,*" I thought as I saw him out of the corner of my eye, tumbling ass over tin cup down the slope. "I've killed the Company's prize engineer!" I always thought in two languages. As soon as I judged it prudent, I let go of the vine and went crashing down the hill, alternately mumbling apologies and cursing the vine. My landing was near where the victim and Matus were.

Soon, Matus had retrieved Mr. Crawford's hat and picked up most of the articles that had dropped out of his pockets: Camels, matches, pencils, and notebook. Mr. Crawford was calmly picking leaves out of his hair and seemed to be all in one piece. He lit another Camel, gave a small laugh and patted me on the shoulder. The engineer said something which I didn't catch, checked the instruments on my belt which came through unharmed, and we set off again. But I did notice that for the rest of the day. Mr. Crawford would turn to look back at me every once in a while. I don't know whether it was to see if I was all right or whether he feared another sneak attack from the rear. I'll always wonder.

Fruit Company engineers lived in virgin rain forests in southwest Costa Rica when surveying began for the Golfito railroad in 1937. (Photo by Clyde Stephens)

The Lottery

By

Billy Saravanja

Picture Golfito in the early 1940's. The new banana division in southwest Costa Rica was being carved out of virgin jungle; engineers were building railroads, bridges, roads, shops, water systems, housing, and airports. Banana farms were being planted and there were all sorts of other activities going on at the same time. Of course, there were deadlines to be met. This, in turn, meant long working hours for everybody.

When development began in Golfito, the area was certainly no French Riviera. There were no telephone communications to San Jose and no local radio stations from the capital reached Golfito. Even electricity was rationed. The place was totally isolated from the rest of the world. The only way out was to risk your life riding in Sam Penry's old airplane to Quepos and then on to San Jose along with crates and mail. Although there were no air conditioners, ceiling fans stirred the hot and humid air. And then there was always rain.

Anyone living in Golfito during those years fully understood the meaning of rain. Rain made the streets muddy and rain made the drinking water from the mountain turn chocolate-colored. Rain caused landslides and washed out the railroad. Rain soaked your feet on the way to work and ruined your good pair of shoes just bought in the commissary.

With all of these conditions, plus the long, hard working hours, it was easy for a person to be driven to misery. That is, unless a person savored the good times of beer-drinking, rolling *chingona* dice for the next round of drinks, making boisterous laughter and enjoying life in the club. But if you were not one to partonrize the club daily, then you were likely to be unhappy.

The Accounting Department had been installed on the second floor of the new administration building. Like all accounting

---

William Kenneth Saravanja was born in 1947 in the Golfito, Costa Rica, hospital which his father, Johnny Saravanja, a Yugoslav, had just built. Billy and his wife, Cathy Gabourel, a Bay Islander from Utila, Honduras, both went to Company banana schools. He is a third generation *bananero* since his British grandfather Bailey worked for United Fruit Company in Guatemala where Billy's mother, Alva, was born. After graduation from Mississippi Southern with a degree in architecture, Billy joined the Engineering Department in Golfito in 1973 and later transferred to Puerto Armuelles, Panama. They now live in Orlando, Florida.

offices in those days, it was filled with dozens of clerks yanking away on their handcrank calculators with miles of white paper strips unraveling out of these machines onto the floor.

Secretaries banged away on heavy old typewriters typing letters with five carbon copies of everything. Office boys carried papers and files back and forth to the many desks all day long. Running this department was Mr. O. E. Beckstrom who sat at his desk, buried in papers and files, and smoked big, long cigars all day long.

Fernando Alvarado, head of the *planilla* — payroll — section, was a middle-aged man, married with a family, and was exceedingly strict about accounting procedures. He drove his staff of clerks to perfection in computing payrolls and closing the books on time every month. Occasionally, it was the opinion of some people that Fernando showed signs of job stress, mixed with a certain misery of the ambient conditions of Golfito. The man also had to put up with some of the smart-aleckness of his junior clerks, some of them Yankees.

On one particular week-day, one of the clerks in *planillas* approached Fernando at his desk for some advice. The clerk noticed that the center drawer of Fernando's desk was partially open and the pieces of lottery purchased by Fernando were clearly visible. The clerk finished his conversation with Fernando and returned to his desk, where he quickly informed the other clerks of the lottery number that Fernando had bought for that week. It was decided among the clerks that it was time to play a little joke on Fernando, and so a plan was connived.

Since Golfito had no communications with the outside world as to what number played in the lottery each Sunday, it was common to wait up to a week before someone got news of the winning number that played the previous Sunday. Normally, the first person to find out what number had played would inform the Golfito telephone operator who would then relay the winning number to anyone who called the central switchboard.

The *planillas* clerks all agreed that when Fernando came around asking what number played last Sunday, they would clue him in on his particular number. In addition, they called the telephone operator and had her agree that if Fernando called her to inquire about the lottery number, she would give him his number, making him think he was a winner.

A few days after that particular Sunday, Fernando came around to his clerks asking if anyone knew what lottery number had played. The clerks tactfully stammered about, giving Fernando only a part of the series that supposedly played, which quickly raised his curiosity. He immediately called the telephone operator, who, according to the plan, gave him the number which lead him to believe that he was a grand winner. Since he had several pieces of the same number, it didn't take Fernando very long to figure out that he was "instantly rich."

Nobody ever realized how Fernando was going to react. Unbelievably, he stomped into Mr. Beckstrom's office and with one sweeping motion of his forearm, wiped all the letters and files off Mr. Beckstrom's desk onto the floor. He then jumped up on Mr. Beckstrom's desk, ranting and raving about what a lousy job he had and what a miserable place Golfito was, and now that he was rich, Mr. Beckstrom could take his accounting job and shove it. Mr. Beckstrom, taken completely by surprise, sat back in his chair, with his cigar sticking out of his mouth, staring up at Fernando. When he was through, Fernando jumped down on the floor and left the office.

The *planillas* clerks were completely shocked by Fernando's behavior. They all ran into Mr. Beckstrom's office and explained the joke they had played on Fernando, and discussed the stress he had been under. The jokers asked Mr. Beckstrom to pardon him. Soon, Fernando found out the truth. After he got over the shock of his life discovering that he had won nothing, there was no alternative left but to go back and ask Mr. Beckstrom for forgiveness. Mr. Beckstrom understood and forgave him.

Fernando remained working in the accounting until his retirement. The Company gave him a big farewell party and the clerks from *planillas* presented him with a full sheet of lottery tickets for the coming Sunday as a going away present.

A Canadian in the Tropics

By

Jean Solovan

First Impressions

In February of 1946, I received the following answer from the United Fruit Company regarding a query about employment: "Tentative offer of employment as a stenographer in the Executive Department of the Guatemala Division at Tiquisate at $150 a month. Transportation and travel costs paid from Edmonton, Canada, to Tiquisate. Sleeping and living quarters free. Six weeks vacation every year after the first eighteen months, with free steamship transportation to the nearest home port and return." Other details followed and I accepted with excitement.

In June, the Company sent an exciting letter outlining my travel arrangements: Leave Edmonton on June 8 by train to Chicago, overnight, pick up pullman ticket at the UFC office for travel to New Orleans, stay at Saint Charles Hotel. Go to United Fruit's office for instructions on getting Guatemalan visa and plane ticket to Guatemala City. Contact Company agent in Guatemala City for travel to Tiquisate.

Fortunately, my trip to the Tropics was made as scheduled. Upon arrival in Guatemala City, I went to the Company office the next morning. An attractive American secretary introduced me to a friendly General Manager, W. L. Taillon, an American, who asked if I wished to stay in Guatemala City for a few days and look around. Thinking that I could be fired for not turning up to work as soon as possible, I hurriedly declined. I was handed a railway pass to leave the following morning. The manager in Tiquisate, Andy L. Bump, would be notified and someone would meet me in Rio Bravo, seven miles from my destination, because that was as far as the train ran. I did not know it then but it was a United Fruit Company train.

The train left on schedule at 8 a.m. I then realized that I did not know the arrival time so I asked the conductor if he spoke

---

Jean Solovan was born about 1910 and grew up in an immigrant farm community in eastern Alberta, Canada, near Vermilion. Both parents were from the Ukraine and Jean grew up steeped in Ukrainian tradition. She translated her father's memoirs from Ukrainian into English and wrote *The Canadians* to record the family history, which included her time in the Tropics with the United Fruit Company. After Jean visited the homeland of her parents in Russia in 1986, she returned to Vancouver. She died in early 1987.

English. He shook his head, "No." I pointed to my watch and said, "Rio Bravo," and his finger pointed at 1 o'clock. I did not understand a single word of the animated conversation among the passengers.

Guatemala City has an elevation of 5,000 feet with a cool, ideal climate. It was getting hotter and hotter as the train continued winding down and around the ever changing scenery of plateaus, valleys, a beautiful large turquoise lake, chalets, thatched roof adobe houses with small Indian children and chickens, and trees loaded with oranges and grapefruit. Lower down, palm trees loaded with coconuts, plots of corn, and other crops began to appear.

About noon, I began to peer out each time the train stopped. The passengers nearby, all men, noticed and would say, "No Rio Bravo." About 1 p.m., several voices sang out, "Rio Bravo," and I heard "Mees Sooleevan" being paged by a young, attractive Guatemalan girl, Marta Mancilla, who spoke a lot of English. The chauffeur was waiting for us in the manager's car.

As we drove along, I didn't see any jungle and was pleasantly surprised when we entered a neat town with nice homes and well kept lawns. There was a group of apartment buildings, elevated about eight feet off the ground on cement posts and painted yellow, the color of ripe bananas. The car stopped at one of these rent-free edifices which I was to share with Marta for a short time.

The big living room was furnished with a rattan couch and chairs, and a writing desk. The coffee table had a big bouquet of powerfully scented gardenias. There was also a Frigidaire for cold drinks and drinking water, and an antique Banana Company telephone for making local calls. The apartment had two bedrooms at the back with a bathroom in between. The downstairs area was a breezeway and served as a laundry facility.

Marta walked to the messhall with me for supper where I was introduced to a very obliging American lady, Myrtle Bryant, in charge of the messhall and apartments.

We started work at 7 a.m., had a two-hour lunch and siesta and then worked from 1 to 4 p.m. I was out of bed at sunrise and went to the messhall for breakfast. The waiter asked me something in Spanish and I shrugged my shoulders indicating that I did not understand. He returned with a stack of pancakes with butter melting over them. I learned later that he asked if I would like *pahncaykees,* so I began to pay attention.

I had reported to work in the manager's office but was asked to help out in the accounting department for a month while their stenographer was on sick leave. One of my jobs was to check the payrolls against personnel cards for deductions of pension and medical plans. I began to get acquainted with names like Diaz, Espinoza, Gutierrez, Lucero, et al. Each name had the father's surname and the mother's initials, or the father's full name and the mother's full name, or, in error, the mother's full name and the father's initials. All that had to be unscrambled. And places such as Heuheutenago, Mazatenango, Retaluleu, Solala, etc. had to be pronounced properly. Employees were allowed to charge goods in the Company commissary and at the club by signing chits. The stack of deductions would be handed to the comptometer operators for payroll deductions.

My first cheque was sizable because of back-pay from April 1, plus travel expenses. The standard deductions were $1 a day for meals, 2% medical coverage, and 4% pension plan. Monthly club dues were $2 which entitled a member to see two movies a week (U.S. films with Spanish subtitles), attend all social events in the club, use the golf course, tennis court, swimming pool, and library. There was no tax.

I had a free, fully furnished apartment, including maid service. Company bedding and towels were taken to the Company laundry for free service, and I hired a laundress for my personal items for the then going wage of $5 a month. She did the washing and ironing in the breezeway downstairs for several others as well, so she did well financially. The woman did not speak English so the first words I looked up in my little dictionary were "to iron" — *planchar.*

The river meandering on the Company property was handy for the village women to do their washing. They stood knee deep in the water and slapped the clothes on river rocks, then spread them on the bank to dry while the children played in the water.

I was there only a few days when a jolt woke me from a deep sleep. My bed was swaying. I thought that I was on the train but the lights were being switched on in the adjoining apartments and there was a lot of excited chatter. Marta came into the room and, by then, I realized that it was an earthquake, which made me feel slightly ill. Marta was very proud of her country and I think I hurt her feelings when I told her that we did not have earthquakes in my country, meaning Edmonton, Canada. At noon I tried to study Spanish and, as usual, dozed off. When my bed began to shake, I flew out of it. Marta was laughing hilariously. She had crouched at the foot of the bed and said, "I make you a little earthquake."

La Despedida

At the end of my first week, I was invited to a farewell party, a *despedida*, at the club for the Superintendent of Agriculture, Walter Burns, and his family who were being transferred to Santa Marta, Colombia. At 5 p.m., the club was full of ladies dressed in light tropical party clothes. The men wore white or light gray suits. Soon the party was in full swing. Marimba players were beating out lively music for dancing, while waiters were passing out highballs. There was White Label scotch, Gordon's gin or Bacardi rum.

Everyone knew each other and I detected strong ties of friendship. A number of ladies introduced themselves, knowing that I was a new employee from Canada, and, in the same breath, asked if I played bridge. They also invited me to come over anytime for a cup of tea or a highball, and pointed with a wave of the hand, "just over there." A young Latin bowed before me to dance. I think all Latins are excellent dancers perhaps because they never sit around when there is music. He suggested he teach me Spanish, I teach him English, and he would come over to my house and bring his guitar. I don't think he expected an answer, as I gathered from Marta, that some Guatemalan young ladies were still being chaperoned in those days.

Fourth of July, 1946

This was an American holiday celebrated in tropical banana zones. First, Marian Hansberger, my new roommate, myself and about 30 others had cocktails at the home of June and Vaughn Kent. Afterwards we went to a dinner dance in the club in formal dress. The club was beautifully decorated with a red and white motif. Myrt saw to it that the cooks outdid themselves with the meal. The tables, extending the length of the messhall, were brightened with colorful hibiscus and loaded with food. The dinner music was supplied by an orchestra brought in from Guatemala City, complete with marimbas. Refreshments, easily available from the club bar, were free for a couple of hours.

Dancing started right after dinner. After tamales and coffee at midnight, dancing continued. There was a custom that as long as anyone danced, the orchestra played on, even until early morning. I loved this Latin music which lasted until 3 p.m. After being dissuaded from going home, I was invited to join a lively group for breakfast at the home of the M and S Superintendent, Mr. Sibley.

The sun was up when I trailed home in my evening gown but no one thought anything of it. This would have been a disgrace back at home. I barely fell asleep when the telephone woke me up. Some of the group I left an hour ago were inviting me to go to the quinine farms. The Superintendent had to attend to some business and thought that I would like to see that part of Guatemala.

The narrow mud road went winding around as we crept up the mountainside in a Jeep. Now and again, we cautiously passed an Indian carrying pottery or wood on his back or leading a mule loaded with wood, waiting patiently on the edge of the narrow road for us to squeeze by. The valley grew deeper and deeper and farther away.

At last, we reached the Company house occupied by an American lady Ph.D. who was there with a staff to oversee the chincona trees, the bark of which produced quinine to combat malaria. The gracious lady invited us to have lunch, and while two servants prepared a meal, we had a cool drink and visited. It had rained in the interim, greasing the road to a treacherous condition. I expect some were saying silent prayers on the way back to Tiquisate.

Community Activities

There was no excuse for anyone to be lonesome. Tropical employees worked hard but they also knew how to enjoy good clean fun. Social life here surpassed anything I ever experienced in Edmonton. We had movies, dances, bridge and tea parties, dinner parties, and a pool. Sometimes we made a trip to the Pacific Ocean's warm water with miles of sandy beaches.

In the Fruit Company zone, everything was within walking distance, including the golf course where there were many excellent players. A tournament was arranged with employees from Honduras who arrived by air in the Company DC-3. After the pros finished their game, many employees were enticed into the act; beginners joined top-notch players. A cart with Guatemalan beer was wheeled around the golf course and a barbecue completed the tournament. Many old friends from other divisions enjoyed getting together again and talked about the good old days working together in other countries. It was like a homecoming.

Christmas in the Tropics

I started my Christmas mail early. To get away from letters for a few minutes, I gazed out the window and was astonished to discover that the leaves on the big bush by the stairs were turning red. It was a poinsettia and conveyed a true Christmas mood. Marian said that Christmas would be one day that she would cry.

Several days before Christmas, young Guatemalans organized a *posada.* Like the Holy Family, they called on several homes to be accepted for the night, each time being turned away, until their arrival at a pre-arranged home where a party was to be held. This continued for a number of evenings.

Marian and I had been invited for Christmas Eve dinner at the home of Guatemalan friends, Pancho and Tina Souza. Both worked in the Accounting Department and spoke English. This was followed by midnight tamales at the British home of George and Gwen Harris, noon dinner on Christmas Day at the American Kents, and back again to the Harris home for Christmas supper. There were seven cocktail parties on Christmas Day. Just in case, the manager's wife, Mrs. Bump, also kindly inquired if I had a place to go on Christmas.

At 4 p.m. on Christmas Eve, the young son of Pancho and Tina brought presents for Marian and me on a tray covered with a lace doily. Our invitation was for 7 p.m. Their home was festive with a Christmas tree and a houseful of doctors with their wives and other Spanish-speaking friends who conversed in their own language — animated talk with lots of laughter. Both Marian and I were happy just being in the merry atmosphere. I finished my cocktail and longed for food — 11 a.m. lunch was a long time ago. Pancho sat beside me and said that I looked *triste.* They wanted me to be happy in Tiquisate. I tried to convince him that I was happy; I did not add that I was also starving to death. The delicious turkey dinner at 9 p.m. was well-worth waiting for and I learned that this was the usual hour for dining.

Guy Harris lived only three houses down the same street and we were leaving to be there for midnight tamales. For some reason, we did not tell our hosts where we were going next. Pancho, like a gentleman, insisted on walking us home across the airstrip which ran through the center of the Company town. After he escorted us to our door, we followed him back to the Harris house feeling silly and giggly.

Christmas Day, now into the dry season, had a clear blue sky with the normal hot atmosphere. We wore sleeveless cotton dresses and sandals. This saved us money on hosiery as nylon was too hot. During this holiday season, some of the cocktail invitations were being fulfilled right after breakfast! Christmas cake, eggnogs, or other cocktails were served — it was very Christmasy. Needless to say, we ran into the same people several times.

Lunch was at the Kents. Someone brought my mail that included numerous cards with snow scenes — an appealing thought in this 99-degree temperature. I tried to visualize my feet in a tub of snow with some being sprinkled on my back. Otherwise, Christmas spirit here was the same as at home, as were some of the meals. This included plum pudding at the Harris home, with rum sauce made by their son Guy, a U.S. graduate engineer. We finished all our social obligations by 11 p.m. and Marian said, "You know what? I forgot to cry!"

My First Vacation

I had been in Guatemala fifteen months and was going home on a six-week paid holiday. The UFC agent in Guatemala City made my travel arrangements: a banana cruise ship from Puerto Barrios to New Orleans, a plane to San Francisco to spend a few days with my sisters, Katherine and Rachelle, and then by train to Edmonton.

I was required to see the U.S. Consul in Guatemala City personally to obtain a transit visa. Also, I would change Guatemalan Quetzales to U.S. dollars and shop. The Pan American Hotel was very handy and granted a 10% discount to United Fruit employees — $12 a day which included all meals with good food and service. The Guatemalan waitresses were dressed in their colorful native dress and served quietly and efficiently. All women wore a lot of jewelry — even small babies had their ears pierced. With so many gifts to take home, I was given a good discount. Guatemalan woven table cloths with matching napkins had a lot of embroidered designs such as chicks and roosters and were almost identical to the type our Ukranian farm neighbor at Slawa drew on Easter eggs. For men, silver cigarette lighter cases were reasonably priced.

Continuing my train trip from Guatemala City to Puerto Barrios on the east coast, I could understand what the passengers were talking about. They were a group of American tourists who had spent a week in the capital city and were also sailing on the same United Fruit Company ship.

As the train pulled onto the Caribbean wharf, there was a lot of activity loading 75,000 bunches of bananas onto a spotless, white, 99-passenger cruise ship. There were six of this type in the Great White Fleet — one named *Fra Berlanga* for the friar who brought banana rootstocks to Santo Domingo from the Canary Islands in 1516 A.D. The luggage was taken on board by Company employees (no tipping). My first sailing was luxurious since no one booked the suite, complete with bathtub and big, fluffy towels. I was lucky enough to occupy it all by myself. There was a very extensive menu. One could dress properly and eat in the dining room or stay by the swimming pool and eat on deck where a table was creaking under a marvelous array of food. Passengers were happy and passed the time at deck games. The captain's dinner party was a delight as we glided smoothly on the Caribbean for three days to New Orleans. What a wonderful vacation. Before I knew it, my six weeks were over and I was back in Tiquisate, Guatemala to begin another year.

Life in Bananera

After some time, I was transferred to the manager's office in Bananera on the Atlantic side of Guatemala. The manager, George Munch, was one of the engineers that had come to the Tropics as a young man to survey lands in the torrid, infested jungles of Central America, all in the name of Chiquita bananas. As a boss, he dictated concise letters. One letter regarded abaca, or Manila hemp, which the Company had been producing as a war material for making rope. Mr. Munch dictated a trip report after inspecting abaca in Panama, Costa Rica, Honduras and Guatemala. This particular letter was long. When there was a backlog of work, I just stayed after closing hours to catch up. But it was easy to put in an extra hour or two because I had enough leisure time. There was maid service to take care of the apartment at Company expense and a laundress to do my personal clothes. I ate meals in the messhall.

Although I brought golf clubs back from vacation, I was a poor competitor for the excellent players, one being Etta Bain whose Scottish father taught her the game. The golf balls which I whacked usually landed in the middle of the lagoon instead of over it. Small Guatemalan caddies capitalized on this by retrieving and selling sunken balls back to the players.

The assistant manager, D. E. Rayor, an American lawyer, also dictated letters and suggested that I learn to operate a teletype. Peggy Baily was going on vacation and I would have the added job for a month. Tropical Radio and Telegraph was Company-owned and all radiograms were received at the port office in Puerto Barrios and transmitted further inland to us in Bananera. The teletype had the same keyboard as a typewriter. Gummed tape was inserted from the right, placed through the machine to the left and into a gadget filled with water. The messages typed at Puerto Barrios came out simultaneously at our end of the circuit in Bananera. They were imprinted on the tape which was then moistened and stuck on a sheet of paper as the radiograms were coming through. The same procedure applied to outgoing messages from our end.

The teletype began clanging the first thing in the morning and I had to sit there and paste. Otherwise, the tape, sticky with humidity, would billow out and stick together, or bunch up in a little sticky heap. Next, radiograms had to be telephoned to all concerned, then confirmation copies were typed for distribution. For a message about a ship's arrival, copies went to the Superintendent of Agriculture, A. C. Hamilton, to arrange the fruit cut, the Division Engineer, Ike Smith, to schedule the trains, and Wharf Superintendent Eddie Taylor to prepare to load. The Manager had copies of all radiograms.

The first Saturday morning after Peggy left on vacation, I dispatched a long radiogram to Boston about production and shipments, hurrying to get all messages sent and received. Then I could get at a pile of letters before work stopped at 1 p.m. There was tapping on the keyboard — a signal to stop. A message came back: "Repeat the entire message, tape stuck." I began repeating and then more tapping started. "Message garbled, type RYS to clear the circuit." I did for awhile, then got exasperated and typed: "I would like to throw this damn machine through the window." The port operator, Rosita Papadopelo, of Greek origin, who spoke English, and I carried on a conversation for a while, typing messages back and forth. Naturally, that part of the tape was destroyed.

The code book was locked in the safe and the safe combination was entrusted to me. The first radiogram I decoded revealed that Guatemala City was under martial law. There had been a political skirmish but this was soon resolved. We were about 150 miles away from the capital.

Sue Munch, the manager's wife, and several other young ladies had signed up with the United Fruit Company for tropical office work as single women during the depression years. They were soon snapped up by young engineers working in Middle America. Sue and George Munch had three children: Suzanne, Sandra, and John and five servants. Sue was an ideal manager's wife — gregarious, imaginative, and fun-loving, creating an atmosphere of relaxation which contributed to enjoyable parties.

I was invited to the manager's home for a dinner party given in honor of a high-echelon visitor from the head office in Boston. The manager, as I knew him, always had his nose to the grindstone during business hours. However, I was surprised to note at the party that he flirted with the ladies and held animated conversation with the men. When George Munch introduced me to the guest of honor as a "Canadian who doesn't do a tap of work," I knew that I had been accepted.

Since some of the employees lived away from the main division headquarters, they depended on the manager to keep them happy and contented. There were parties at the Golf Club, but a gathering in the manager's home to meet important people or observe special occasions was of special significance as a morale booster. As the guests arrived, there were always affectionate greetings and happy conversations, but the men invariably drifted off into a corner to talk bananas. Frequently, the ladies tried to disuade them from "talking shop," but I observed that many problems were taken care of at these social functions. Perhaps my thinking was somewhat biased because I was keenly interested in the Company operations more than most wives were.

Naturally, there were jealousies and rivalries. In cases of promotions, sometimes there were preferences for friends. A few arguments arose now and again which the manager had to smooth out until Labor Relations could handle such cases.

Archaeology

The United Fruit Company was a great contributor of funds and talent used for the restoration of archaeological sites. Upon completion, these projects, including museums, were turned over to the Guatemalan people who took pride in their Mayan heritage. For guidance, the UFC had advisors from the Carnegie Institute and other organizations.

Among the numerous remains of the ancient Mayan cities buried in the jungle for hundreds of years, Quirigua, only a few miles from Bananera, was one of the most interesting. Here, some of the finest examples of Mayan culture and art were brought to light. The site became known to the world in 1840 by two men, Stephens and Catherwood, who were interested in archaeology and traveled through Central America. Also in the 19th century, an English archaeologist spent several months at the site. It was not until 1910 that extensive excavations were made.

One of 14 monoliths visited by pioneer banana people developing plantations around the ancient Mayan city of Quirigua in the Motagua Valley of eastern Guatemala. Artifacts date from 400 B.C. to 900 A.D. During the early 1900's, United Fruit Company cooperated with archaeologists in the restoration of temples, courts and fallen monoliths and preserved the tropical forest between the ruins and new banana farms. Photo taken about 1910. (From the files of United Fruit Company)

United Fruit acquired title to the Quirigua site in 1909. The following year, adjacent lands were cleared and planted with bananas, but a thirty-acre tract of tropical forest surrounding the ruins remained untouched. At the time, an expedition from the United States undertook a series of investigations and two of the principal edifices were reconstructed with the aid of the United Fruit Company.

No further excavation of any importance was made until 1934 when men from the Carnegie Institution, at the request of and in cooperation with United Fruit, re-erected the fallen stelaes. The monoliths varied in height from ten to thirty feet, one weighing about 65 tons. Great care was taken in lifting the enormously heavy, sculptured monuments. The human figures, not unlike the present Guatemalan Indians, were sculptured on both sides of the stones and were shown with magnificent head dress and elaborate clothing. These figures represented Mayan ruler-priests who were also great chronologers and astronomers. The remaining two sides of the stones were covered with columns of hieroglyphs. The ones that have been deciphered refer to the recording of time. An unexpected discovery was the finding of two great altar slabs. Having been covered by silt, the carvings were in excellent condition.

The Quirigua area was made into a park. With the railway track nearby, guided tours were arranged for banana ship cruise passengers who did not wish to go to Guatemala City while the Company ship was in Puerto Barrios being loaded with bananas. After a tour of the ruins, passengers returned to Bananera for lunch and shopping or a swim in the Company pool.

A lot of excitement was generated for a few days when actor Ralph Bellamy and his wife were guests in the manager's home. Bellamy and his wife were also interested in archaeology. The photographer Rye and his wife took some excellent pictures of them viewing the monoliths in the Quirigua plaza. I coaxed the Chief Clerk into letting me deliver a radiogram to Ralph Bellamy in the club and was rewarded with a firm handshake from one of my favorite actors.

The Flood

Occasionally, there were bad tropical storms. Sudden winds drove rain through the screen windows into the apartment and soaked everything before awnings could be lowered. Thunder and lightning illuminated the landscape and branches snapped off trees. The next morning was calm and sunny but there was a lot of debris everywhere for the yard men to clean up. These "blowdowns" also caused heavy losses in the plantations because bananas blew to the ground.

The fall of 1949 was the worst flooding in the country's history. Usually, mid-September was a period of heavy rains averaging about 15 inches in the Pacific lowlands. That year, rainfall in Tiquisate measured 25 inches in September with most of it falling the last few days of the month. This was converted into a flood creating a trail of devastation. Even highland lakes overflowed their banks inundating mountain villages.

The Pacific lowlands were affected the worst as the water ran down the slopes flooding the fields and wiping out roads. A mountainside collapsed sending rocks and earth hurtling into the farms and villages. Hundreds of lives were lost and thousands were left homeless and without possessions. The crops were plastered to the ground.

The news of the Guatemalan flood reached Canada and my worried family sent a radiogram inquiring about my safety. However, Bananera, where I now lived on the Caribbean side, had been experiencing only steady rain. Nevertheless, by mid-October, the Bananera farms were also inundated and overseers were getting around the farms in *cayucos* — dugouts. The water in the Bananera lagoon began rising and covered the bridge that connected some of the homes. We watched apprehensively, but fortunately the water receded before much damage was done.

When the rains, stopped there was much work to be done. Clothing, food and money were collected for the homeless. United Fruit provided corn seed for replanting and medicine for the control and prevention of diseases in many devasted areas.

During this disaster, it was eerily quiet — there were no trains, no planes, no mail, no anything for a number of weeks.

The Election

There was an election one weekend and people had time off for voting. It was quiet everywhere. I was invited to go on a launch trip to Rio Dulce with my friends, the Bains and Scotty and Millicent Crichton. Our plans were to board the afternoon Company train (free) to Puerto Barrios, spend the night in the Company hotel (free) and leave in the early morning on a free

Company launch. The launch was complete with captain and crew and was maintained by the Company for the use of employees for sightseeing and fishing trips. When we arrived in Puerto Barrios by train, there was a lot of scurrying around. When the Superintendent saw us, he shouted, "Go on board the ship immediately. There is fear of shooting and we are evacuating women and children to the ship for safety." When we hurriedly boarded the 13-passenger ship, I began worrying about food, but was assured that there would be enough for two days to feed the 100 or so now on board. The Company ship would sail away in case of trouble.

There were rumors circulating that a cache of ammunition was found and that there had been some shooting. Darkness fell. Etta and I were standing on deck and thought we heard a couple of shots in the distance.

We were still in port at midnight when we were assigned to the ship's hospital to sleep. The next morning there was some loud pounding on the door and Scotty was shouting, "Get out of bed right away; we're going on the launch trip." Apparently things had quieted down. We boarded the launch and enjoyed the fascinating scenery of Rio Dulce and a gorge connecting Lake Izabal with the Caribbean Sea.

Returning to Bananera that evening, everyone was quiet. Only the train was making a clackity noise.

A Second Chance

When George Munch was transferred from Guatemala to Almirante, Panama, in 1952, there was a big round of farewell parties for him and Sue. Even the office boy was upset and wrote a note saying that he could not say goodbye because tears would be streaming down his face. At the end of the note, the office boy, a Guatemalan Indian, added a P.S. which said, "How about a little increase in salary before you go?"

For various reasons, I decided to quit the Company and returned to Edmonton, Canada, in mid-1952. It didn't take long for me to get over my home-sickness and ask to return to the Tropics. I was re-employed again in Tiquisate and arrived back in the Tropics from New York aboard the *Yaque*. A year later, George Munch offered me a job in his manager's office in Panama which I accepted.

The Bocas Division on the Caribbean side of Panama had been out of bananas because of Panama disease since the 1930's.

Old banana lands had been planted to cacao. During World War II, Changuinola, the farm area, was planted with abaca to make rope for the Allied navies and armies. Only recently, 2,000 acres of Gros Michel bananas had been planted on flood-fallowed soil and plans were made to expand production to 20,000 acres. This was in addition to the present acreage already in abaca and cacao. All of this meant a lot of new activity for this old division and a lot of work had to be done.

Transfer to Almirante, Panama

After flying from Guatemala, I changed planes in San Jose, Costa Rica, for the flight to the province of Bocas del Toro, Panama. In a short time, we were descending in Changuinola right in the middle of the beautiful Chiquita banana farms with the Costa Rican mountains still in view. It was like coming home — Marian Hansberger Bishop was waiting for me at the airfield. She and her husband had been transferred from Guatemala to Costa Rica and then to Panama where "Bish" was division engineer. Marian taught school but took the day off to accompany me to the main headquarters at the port of Almirante, about 15 miles away by rail.

A motorcar, a regular car equipped with steel wheels to run on rails, was nearby. The motorcar operator put my belongings in and waited for me to have lunch in the Bishop home, a short walking distance away. "Bish" joined us for a highball while the servants put the finishing touches on the meal.

Changuinola consisted of homes for the division engineer, superintendent of agriculture and their sizable staff, office buildings, a school, church, club, and an automotive shop to maintain trucks, tractors, and draglines. All of this was scattered along the tracks from Base Line near the airport and abaca plant to Farm Eight and Empalme.

Marian and I had much to talk about. As we chatted aboard the motorcar enroute to Almirante, I glanced now and again at the green jungle, exotic yellow blooms of the bird of paradise, white ginger flowers, masses of vines and thickets of trees. There were no roads; all traffic moved by rail, either by train or motorcars. The motorcar operator stopped periodically to check by telephone with the dispatcher to assure that the rail line was clear.

After an hour or so, we arrived in Almirante and soon passed the machine shop, the Material and Supply building, the commissary, the police station, and numerous Company houses. I fell in love with this banana port the minute I saw the placid, picturesque bay. The calm harbor was dotted with small islands in the distance and the Caribbean opened up beyond Almirante Bay.

The rail tracks extended near the office, a big three-story concrete edifice a few paces from the sea wall. The motorcar operator carried my luggage to a two-story house a few feet away while Marian and I went to the office to see the Manager. George Munch's greeting was typical: "Where the hell have you been? There is work to be done here!" But he had a big smile on his face.

Christmas in Almirante

The parties began early in many homes. As usual, Mike and Doris Campbell had an open house party. Guests were welcome at any time to sing if they wished while Doris played the piano. The bachelor boys arranged a successful party with Bob Bruce, chief clerk in the manager's office, who contributed with his social experiences learned during his time at the U.S. embassies. Bob Richards, the accountant, and wife Betty, hosted a big cocktail party in the office building. Then employees were invited to the home of Tom and Phyllis Eckersley out on the line in Changuinola. He had been employed in Peru with a railroad company before joining our subsidiary, Chiriqui Land Company, in Panama.

I bought a Canadian Christmas tree for the exhorbitant sum of $4.50. Back in Canada, they sold for 25 to 50 cents at that time, but it was expensive to bring them all the way to the Tropics. The tree was small and flat, having been stacked with other trees, but after soaking in a tub of water, it came to life again. The fragrance that the tropical heat brought out was well worth the price.

Celso "Mac" Macari, a young American traveling auditor, came over and overcame his homesickness somewhat by decorating the tree with the assistance of my Canadian roommate, Maggie Guy. We also made candle holders out of liquor bottles. Ribbons and wrapping paper were not easily available so what was brought back from vacations was ironed and used several times. In the meantime, I made chocolate fudge.

To pay off the many social obligations, Maggie and I had a combined Christmas party and house warming which we celebrated on Canadian Boxing Day at 3 p.m. We sent out 60 invitations and stocked up on Panamanian beer, Carta Vieja rum,

Dewar's White Label scotch, and Gordon's gin. A houseboy was hired to chop ice from a big block and serve drinks. An arrangement was made with the messhall to prepare and deliver cocktail food. Because of their excellent cuisine and artistic arrangement, this service was in great demand.

There were no snobs and I think all invitations were accepted. By 2 a.m., all the food was gone, including the fudge. I was happy and pleased and finally crawled into bed at 4 a.m. This is when I discovered masses of nutshells which some joker had placed between the bed sheets. Just as I dozed off to sleep, the alarm clock began ringing — another trick of some prankster. But the worst part was trying to figure out why it was necessary to be in the office before 7 a.m. It seems that some employee from the hospital had to bring a jug of water to the manager's office before 7 a.m. for dispatch to a Panama City water lab where it would be routinely tested for potability. The sample was to be picked up by a messenger from the port office before 7 a.m. and put on the morning launch to the island of Bocas del Toro to connect with the plane destined for Panama City. He could have taken it directly to the port office without bothering me on that difficult day, but it was my duty.

Up at 6 a.m., my roommate came out of her bedroom, saying it was the first time she ever slept in a bar, referring to our place, now full of empty bottles. We shook hands and congratulated each other on a good party. The place had been warmed nicely and happily but was in shambles. By lunch time our little Panamnian maid had the place in perfect order. She was honest and obliging, but there were slight accidents now and again. A note would be left saying: "You ashtray fly out my hand and break" and offer to replace it. Of course, it was not necessary; she had two little girls to take care of.

Recreation in Bocas del Toro

The Company owned several launches for transporting employees, government officials, and others, all free of charge, to and from the provincial capital city of Bocas del Toro on Columbus Island where planes landed from Panama City. (Later, the Changuinola airstrip was opened again.) On Sundays and holidays, a free launch with crew was provided to the employees who wished to go on an outing. A popular spot was Bocas del Drago on Columbus Island about 15 miles away. There was a white sandy beach, warm calm water, and beautiful coral formations. The reefs had fascinating purple sea fans and thousands of small fish darting around the corals. There were also many varieties of conch shells. Several families lived on that end of the island and ate conch meat. These beautiful empty shells were thrown back into the sea, but the prettier ones were taken back to our home for display.

Walking along the beach one day, Rosalie Bayne and I spied a piece of pottery sticking out of the bank. We began digging with sticks. I got a sizeable piece that had a replica of a fish on it. It just happened that a visiting American archeologist and his wife were on the picnic that day. He became very excited about my find, analyzing that it was a pre-Columbian piece and offered to buy it. I gave the artifact to him since it did not have the same value to me. The rumor got out about our dig and a stampede was created. Some of the populace hired a special boat to go there on Monday and the entire area was dug up, producing only broken pottery. Most people only had gold in mind.

Medical superintendent and surgeon, Doctor Gustav Engler, an Austrian who had escaped the Nazis, owned a sailboat, as did superintendent of Materials and Supplies, Tom Dunn. Both of them took guests on leisurely trips in their sailboats, but often they joined the rest of the employees on the launch. Dr. Engler and Tom Dunn were the only two owners of air tanks and were able to dive many feet down, while most of the others amused themselves in shallow water snorkeling. Frequently, someone would come up with a big, succulent lobster and dunk it in boiling water that the island inhabitants had ready on the beach. At the end of the day, there were always happy red faces returning with many items collected on the beach.

We also went in the opposite direction for a weekend outing. There was a train waiting on the track on Saturday noon for anyone who wished to go to Changuinola to play golf, swim in the pool, or sit at the bar. At times, we enjoyed a barbecue with an orchestra. After dancing and visiting friends, we returned that evening to see a movie in the Almirante Club.

For a change of tempo, a group of us were invited to have dinner with the crew on an offshore oil drilling rig. We arrived by Company launch and were hoisted up to an air-conditioned dining room for an excellent meal. This was followed by an orientation tour to watch the drill being changed. Back in Almirante at the club wharf, we had tea aboard a yacht belonging to a retired Bethlehem Steel executive and his wife.

Work and Play

Labor relations were good. There were a number of Jamaican employees and Raymond "Blondy" Gronblad could imitate their lingo when talking with them, much to their amusement. When bananas were being harvested, the manager, assistant manager Ed Spence and superintendent of agriculture, Dudley Linton, put on their banana latex-stained khaki clothes and spent the day on the farms with the workers to ensure that only first class bunches of bananas were loaded on the trains. They returned tired at the end of the day but still joined port superintendent, Yantis Holliday, to inspect the loading of the ship.

Regardless of where the bosses went, Rose, the West Indian Panamanian telephone operator, could always track them down. She knew who went where and was invaluable in locating anyone who needed to be found. Rose was also chatty and cheerful as well.

American Alice Louise Dunn, the hospital administrative assistant, solicited anything that could be used by patients interned in the hospital and needed clothing. There was a full-time seamstress on the Company payroll to sew or alter donated clothing. Alice Louise was particularly delighted one day when English Jamaican Dudley Linton, superintendent of agriculture, finally gave up his red shirt which was made into a pretty dress for a little Indian girl.

Panamanian Independence Day, November 3rd, was celebrated with much fervor with flags flying everywhere and highlighted with parades and dances. But even during the rest of the year, there was always a strong respect for the flag. This was known by the office employees who passed the police station on their way to work everyday. If it happened to be 7 a.m., the National Guard blew a whistle and everyone stopped in their tracks as the flag was being raised. When he blew the whistle again, we continued walking. The same respect was observed in the evening when the flag was lowered. One Sunday, two foreign couples were meandering by the messhall for supper and were talking as the flag went down. The *Guardia* blew the whistle on them and the manager had to bail the pair out. They claimed they did not hear the whistle, which was hard to believe. I thought of the lukewarm attitude of some Canadians and their lack of pride in displaying the flag, even on July 1st.

To enter the messhall for meals, we had to go through the club and pass the bar and a piano. Frank Ayala, a Latin employee from one of *the* families of Panama, was an accomplished pianist and I asked him to play "Because of You," one of my favorite songs at the time. Later, I was awakened at an early hour one Sunday morning by whispering on the stairs, followed by a soft harmony of singing "Because of You" and accompanied by a guitar. They moved away quietly after completing the verse and then serenaded the girls in the other three apartments. This was a charming custom which I'll never forget.

Another Transfer

The year Fidel Castro took over United Fruit Company operations in Cuba in 1959, a number of transfers were made in Central America. George Munch was sent to Honduras, the Bishops were off to Guatemala, Engineer Ike Smith and Edith came from Guatemala to Changuinola and assistant manager, "Blondy" Gronblad, transferred to San Jose, Costa Rica. With someone like Bob Bruce running the Almirante manager's office, it was easy for Guy Cantrell to take over management after his transfer from Honduras.

There were rumors about me being transferred. The satisfaction of being in one place for seven years was diminishing and I was about ready to leave. News of my transfer finally came and I flew across the mountain to Puerto Armuelles on the Pacific side of Panama to my new job in the manager's office. It was nice working with the manager because I had known Aaron and Bonnie King from Tiquisate days in Guatemala. The chief clerk, Franz Eduarte, a Costa Rican, was easygoing and was also nice to work with.

Armuelles was quite different for me. My job was not quite as interesting as the one I had in Almirante, but there was the consolation of having a brand new cozy house, fully furnished and with free maid service. Now I could at last cook my own food in my own kitchen and not bother about going three times a day to the Company messhall. I did miss all the nice vegetation in Almirante but there were lots of birds to enjoy in Armuelles. Another attraction was the large number of big iguanas running all over the place.

Up until now, the banana zones in Panama had been reasonably peaceful. However, the mood changed when communist Cuba's influence spread throughout Central America. In 1960, strike fever spread from Almirante to Armuelles. One morning the usual work whistle did not blow. It was eerily quiet, but I was going to the office anyway and took an umbrella in case it rained. A young man with a machete stepped in front of me and asked in Spanish where I was going. I told him. He said the office was closed and there was a strike on. I could not argue with him — not with my umbrella against his machete — and went home. There were threats to burn down the houses of the foreigners, plus other forms of violence. The illegal strike terminated when about eighty Panama National Guard troops were flown in.

After a vacation in Edmonton, I returned to Armuelles to discover that I had been transferred back to Almirante. It was good to go back to old friends again but the labor strike had tarnished the pleasantness of the atmosphere. This state of affairs made it easier for me to resign in 1962 after sixteen years with the United Fruit Company. On my way back to Canada, I stopped off to visit Doris and Mike Campbell who had retird in California. After arriving at my new home in Vancouver, it took some time to adjust.

Once again, I was employed by the United Fruit Company for the third time. It looked like working for this Banana Company was becoming a habit. However, the last time, I worked only in the temperate zone, indeed a far cry from the Tropics.

Bluefields, Nicaragua

By

Mark Trafton, Jr.

During my first 20 years in the Company, I had heard the story that the worst thing that could happen to an employee, aside from getting fired, was to be transferred to Bluefields, Nicaragua. In April, 1946, I was ordered by a benign boss, Doc Dunlap, to go to Bluefields. The purpose was to make a soil survey of United Fruit Company properties along the Escondido and Siquia rivers to determine if they might be suitable for small-scale African oil palm plantings.

Bluefields, a small Nicaraguan port on the Caribbean, had been the headquarters for Sam Zemurray's Cuyamel Fruit Company when they planted bananas on the narrow flood plains of the Escondido and Siquia rivers. The project soon failed due to poor soils and floods but Zemurray's company retained title to all the land, some 20,000 acres. Native growers were allowed to harvest what was left of the bananas and sell them to coastwise schooners. In 1930, when Zemurray sold out, these lands became part of the United Fruit Company.

Bluefields was also the port where the United States Marines landed in Nicaragua in the early 1920's to suppress the Sandino revolution, leaving a benevolent dictator, "Tacho" Somoza, as president of the country. Many marines returned and stayed on for life in important government positions which Somoza provided for them. I met one of them, Captain Turner, a typical southern gentleman, who was the head customs officer for the port.

I had often been though Managua, the capital of Nicaragua, on flights from Honduras to other Company divisions. The scenery is spectacular, passing the stark 4,200-foot-high Momotombo volcano puffing bursts of smoke at intervals, and then approaching the airfield alongside beautiful Lake Managua, not far from the Pacific coast.

The only certain access to Bluefields, about 200 miles east of Managua, was by air. Overland, after a short automobile trip on a paved road, a four-wheel-drive jeep was required to transit the dry season road through the largest single block of rain forest remaining in Central America. On the jeep trip, one needed camping gear and tents since arrival at the river port of Rama was uncertain and might take several days, depending on the level of water in the many streams. There were no bridges and many jeep expeditions were forced to return to Managua and await more favorable weather conditions. Most of the area is low and has often been considered as the location for a second transisthmian canal.

The flight from Managua was in a small two-motor cargo plane but a few seats could be installed for up to six passengers. On my first flight, the only other passenger was an elderly, well-dressed lady whose daughter drove her to the airport in an expensive car. I chatted with the *señora* who told me she owned a coffee farm in the mountains of the Huapi range to the north and that the plane would make a special landing to let her off there. I wondered about that but, sure enough, on a small grass strip which looked like just a crack in the forest from the air, the plane landed and she disembarked. There was nobody there except a *mozo* who was holding the tether to a large ox with a makeshift wooden saddle on its back. He helped the lady onto the ox, tied her suitcase beside her, and as she waved *adios,* he started leading the ox and its passenger through the jungle up a rough mountain trail. I couldn't believe my eyes. I have often thought of that wonderful woman who so nonchalantly had travelled that morning by automobile, airplane, and oxback from a city residence to a farm in the jungle, obviously considering it a routine journey and nothing to get excited about.

The plane landed in Bluefields about an hour later on a short, muddy, red clay landing strip. I was met by Mr. Romhild, the Company agent, and taken by automobile, one of four in the town, to the "hotel."

Bluefields was a one-street town, about four blocks long, with ramshackle wooden buildings, including four churches, on both sides of the muddy, unpaved street. Houses on the ocean side had their outhouses built over the water with access by a short wooden walkway. Most of the inhabitants were blacks who survived by fishing. Romhild was a Dane who had jumped, or missed, a ship there many years before. His job, assisted by the only other monthly employee, the accountant, was to try to prevent squatters from taking over the Company lands along the Escondido and Sequia rivers, pay taxes once a year, and read the mail.

The Company supplied him with a fancy cabin cruiser which was his pet. He used it mostly for fishing and rarely for trips up the river. The hotel, at one end of town, was terrible but had a large lounge which was the social center for the few local officials, travellers, and crews of fishing trawlers and coastwise schooners. The black townspeople spent the evenings in their several churches praying and chanting. The customers in the hotel bar often turned off the juke box so they could listen to the strange blends of chords and discords emanating from the churches.

The Escondido River and its many tributaries was subject to extreme flooding, especially in the upper reaches where trash could be seen hanging in riverine trees forty feet above normal water level. There were scattered cattle ranches along the wide lower valley and delta area where bananas had died out from Panama disease. Large coastal schooners, trawlers, and small ocean-going vessels could navigate up the Escondido some eighty kilometers to the river port of Rama, a larger and more active town than Bluefields. It was the trading center for the thousands of mestizos who lived in small clearings along the endless miles of rivers and streams in the vast rain forest.

On my second day in Bluefields, Romhild put me on a large motorized steel freight barge, which had a few wooden rooms on the deck, for the 24-hour trip up river to Rama. He said he couldn't take me in his cabin cruiser due to trees and trash floating down the river. In Rama, I was met by the Company land agent who took me each day in his outboard *cayuco* — dugout — to the areas where I would collect soil samples and return at night to the "hotel" in Rama. He also hired two peones to chop lines through the bush in a half-kilometer grid pattern, and to pull the soil auger from a depth of three feet at each station. I didn't have the back for that kind of exercise and limited my labor to labelling the small cloth bag for each soil sample, noting grid location and soil type at depths of twelve, twenty-four and thirty-six inches for each boring. After five hot and rainy days I returned to Bluefields, packed the soil samples for shipment by air to the lab, and flew back to La Lima, Honduras. As a result of the soil survey, four, one hundred-acre experimental sites were selected.

During the next three years, nurseries were established, field plantings made, and, after four years, when the palms commenced to bear fruit, a small processing plant was set up. The mill was located at the down river plot so that harvested bunches from the three other sites up river could be floated down to the mill in *cayucos*. In the meantime, I had approved construction of an overseer's house at the mill site so that I would have a decent place to stay during visits there. A young Nicaraguan was contracted to operate the project until the palms commenced to produce, at which time he took over on his own. It was a successful experiment and the operator made a very good living since the palm oil and kernels were in demand locally at high prices. We had originally planned to put in a large palm oil project after observing results of the experimental plantings, but plans were cancelled due to changes in Company policy and political unrest in Nicaragua.

This had been one of my "babies" from the start and I visited it many times as it developed. Romhild eventually condescended to take me up the river in the cabin cruiser and to bring me back down, provided the Escondido wasn't in flood stage. I think he finally realized that something was going on in his division and that he had better cooperate if he wanted to keep his job. So conditions improved slowly, making my visits there much more interesting and comfortable.

I guess it was true that I had been told about Bluefields. On one trip there, I became very ill with amoebic dysentery from eating oysters which apparently had been kept in a water pen on "outhouse row." Aside from that episode, I actually enjoyed being a "bushman" again and met many interesting people on the farms and ranches along the river, and in the cities, including Managua, where I had contract dealings with government officials.

There are always some pleasant surprises in this life, and one happened to me in, of all places, Bluefields. I had just arrived there one day and was sitting in the hotel lounge that evening with Romhild. We were only on our first drink — but I had a vision! Three beautiful, young, blond American girls, in scanty attire (which was uncommon then), and three guys appeared in the doorway of this dilapidated bar.

They entered, sat at the table beside us, ordered drinks and started singing. The song — "Ghost Riders in the Sky." One of the male singers was Vaughn Monroe, who had recently recorded that very popular hit record. The other two men owned a large seafood company in California and had a small fleet of trawlers and a large refrigerated mother ship operating off Bluefields. They had come ashore to see what night life was like in Bluefields! It wasn't a vision, or a dream; they were there beside Romhild and me!

It was a long evening as we entertained them with drinks and tales of jungle life. My trip up river the next day was postponed while *they* entertained *us* aboard the large mother ship — a floating palace, which just happened to also have refrigerated space for tons of lobster, shrimp, and fish.

Bluefields. Perhaps it had been a possible United Fruit Company "penal colony," and it is undoubtedly worse than that now, but for several years I actually enjoyed, and looked forward to, my visits there.

El Paseo

By

Scotty and Millicent Crichton

In 1947, while living in Los Andes District, Guatemala, we arranged a trip up the Rio Dulce with the Bains, Jean Solovan, and Russ Winters. This was to be a special treat for Millicent's father who had never been out of the States. We were preparing to leave home when our local paymaster rushed up and insisted that we take the Company money out of the safe and transfer it to Bananera. He gave no reason for doing this but we were to learn why the following day.

Our train ride from Entre Rios to Puerto Barrios was uneventful but on reaching our destination we found the town deserted. Not a soul in sight! We finally located a very excited Mr. Papadopolo who shouted, "Get down to the ship immediately!" We rushed to the docks, which were being guarded by local army units, and found all the Barrios residents already aboard the *Morazan.* Seemingly, a freight car loaded with guns and ammunition had been located on a railroad siding outside of Barrios so the authorities believed a revolution was imminent. Everyone had been advised to vacate the town.

Soon after boarding, the *Morazan* sailed out into the bay where we remained for the night, together with all the folks from Barrios. What a picnic it was! There was plenty of food, much attention from the ship's officers, and good companionship.

This was a twelve-passenger ship so sleeping accommodations were extremely limited. The women in our group were fortunate to be able to spend the night in the ship's hospital. Scotty and Willie Bain tried to share a bunk, head to feet, but Willie was a restless sleeper. Scotty gave up and tried sleeping on a bench on deck. After a fitful night's rest, we awoke to find that

---

Henry M. Crichton was born in 1908 in Edinburgh, Scotland. After receiving a degree in electrical engineering at Herriot Watt University, he went to New York in 1929. In 1932, Scotty joined the United Fruit Company as an engineer aboard the old *Tivives.* Later, he was involved in the development of Company divisions at Golfito, Costa Rica, and Tiquisate and Bananera, Guatemala. Scotty and his wife, Millicent, a New Yorker, also lived in Honduras, Panama, Colombia, and the Dominican Republic. Scotty developed and built the cableway hanging tractor to haul banana bunches to the packing stations. The couple retired in 1973 to Fort Myers, Florida. Together, the Crichtons — David, Sr., Dave, Scotty and Millicent — contributed over 120 years of work to the United Fruit Company.

everything was still quiet. We disembarked and made our trip up the Rio Dulce, a bit anticlimatic after the excitement of the previous night.

Upon returning to Puerto Barrios, we learned that the rebels, along with our paymaster, had gone elsewhere. The paymaster was one of the rebel leaders but was apparently a true Fruit Company man at heart. That was the reason he gave the money to us to make sure it reached Bananera safely.

How Quepos was named (according to a UFC Scotsman)

When passing along the Pacific coast of Costa Rica in the 1930's, a storm wrecked the Fruit Company launch. A Scotsman managed to swim ashore and immediately offered a prayer: "Praise God, the good Lord has blessed and *kep us.*" Thus, the new banana division at that site was named *Quepos.*

An Engineer's Logbook

By

Frank C. Sanders

The Jaguar

In 1927, one of our early overseers in the Guatemala Division, Floyd Avery, was such an avid hunter that most of the game disappeared in the "new land" farm area of Los Andes District where he worked. He also trapped and tried to tame some of the tropical animals that he kept.

During this time, a jaguar was known to prowl around their house at night. Sometimes it even entered the area beneath the house. Floyd set a trap to try to catch the animal, but nothing happened for some time. Finally, one night he heard a terrific thrashing about under the house. He took a flashlight to investigate and found a huge jaguar in the trap. From the frantic movements of the big cat, it could be seen that he might escape at any moment. Avery got a long piece of 3″ pipe and placed it across the neck of the animal. Only when he stood on the pipe could he subdue it.

Avery called his wife for help and instructed her to stand on the pipe while he went to get some rope. Because he only had one light, he had to leave the scared woman in the dark while he searched for the rope. While he was gone, she had to hang on to the joists above her head to keep from being thrown from the pipe by the furious, thrashing beast. Floyd returned with the rope and trussed the animal up with great difficulty. Somehow he managed to get him in a cage. Because there was no chance of taming a full-grown jaguar, Floyd gave it to the Aurora Zoo in Guatemala City.

The Prankster

About 1958, a young field engineer arrived from the States to work in La Lima, Honduras. This boy had a phobia about snakes. Most of us had a healthy respect for the creatures, but

---

Frank C. Sanders was born in 1903 in Conneaut, Ohio, and came on the M. S. *La Marea* to Puerto Barrios, Guatemala in 1929. He worked in the Engineering Department and had assignments in Tiquisate and Bananera in Guatamala, Puerto Castilla of the old Truxillo Railroad Company in Honduras, and the Almirante Division of Panama, with brief tours of duty in Costa Rica and Colombia. Mr. Sanders retired in 1968 and settled in Bradenton, Florida.

this boy's fear was irrational. Sometimes he would go out with his *cuadrilla* on the motorcar and get off at a certain place to work in the bush. At this point, he would give instructions to the gang about how to carry on with the work, but he remained along the railroad. He could not force himself to enter the jungle because of his fear of snakes. When the men finished the day's work, they would join him back at the railroad for the return to La Lima. Naturally, this behavior became common knowledge and it made the poor guy a target for super-prankster Paul Vegas.

Paul was a former football player for Tulane University and was more than a physical match for an ordinary person. Consequently, Paul was immune to retaliation for his practical jokes.

To pull a prank on the new engineer, Paul captured a five-foot boa constrictor. Although a small boa is harmless, it is horrendous to look at. At a quick glance, this reptile can easily be mistaken for a *fer-de-lance.* Paul gained entrance to the young engineer's room and deposited the snake among the shirts in a bureau drawer. When the boy opened the drawer to take out a shirt, he was confronted by a horror. All reason left him as he took out his pistol and proceeded to empty it into the closed drawer. He killed the snake, but he also ruined the bureau, riddled the shirts, and made a pepper box out of the side of the building. Luckily nobody was in the line of fire.

I asked Bob Mixon, the Division Engineer, about this incident and he said that he had confronted Paul about it. Mixon told Paul that he was not going to ask whether or not he was responsible, but to please never do it again.

When Paul was transferred to the Bananera Division in Guatemala, his reputation preceded him. One Sunday morning he was present when the veterinarian was called to dock the tails of eleven thoroughbred puppies owned by Sonny Ritter. Paul at once saw an opportunity to indulge his penchant. He retrieved the eleven puppy tails and took them to the hotel, where he induced one of the waiters to make a large platter of salad consisting of lettuce and tomatoes, tastefully arranged around the tails.

This was first presented to Virginia Adams, our lady draftsperson. She stood up and screamed, declaring that the lack of a gun was all that saved Paul's life. The next move was a mistake. The "delicacy" was presented to Armando Martinez, who was the manager's secretary. Armando promptly put an end to the game and the dog tail salad was thrown away after only two pranks.

The Gold Brick

In 1957, Harlo Van Wald was Manager of the Bananera Division in Guatemala. The Guatemala City office of the Company was headed by Lazarus "Hank" Greenberg, who held the title of vice president. The street door was always open during office hours. This door gave access to a patio from which the several offices connected.

One day Von Wald was in the city for a consultation with Greenberg when a barefooted Indian wandered through the street door and into Greenberg's office. There he displayed an object which he purported to have found in the street. The object was about three inches long and three quarters of an inch square. It was wrapped in white paper and secured with sealing wax. Printing on the paper indicated that it contained gold, 997 fine, and that it came from a mining company up in Mexico. It was addressed to a dentist in Belice.

Von Wald and Greenberg volunteered to see that the package would be delivered to its proper destination. The old Indian asked if he did not merit some reward for finding it. The two decided that the dentist would be happy to reimburse them if they gave the Indian a reward of five dollars.

After the finder had departed with his reward, they began to feel that something was not quite kosher. They broke the seal and opened the package to discover that it contained a bar of brass. Everybody agreed that this Indian had talents. Von Wald kept the wrapper as a souvenir.

The Concrete Mixer

We were erecting a 125-foot steel railroad bridge across the Rio Siguacan at the entrance to Isabal Farm in the Tiquisate Division in Guatemala. This was about 1937 and I was the Railroad Engineer (not to be confused with a locomotive engineer). The concrete abutments of the bridge had been poured and the construction gang was busy with the false work preparatory to the erection of the steel. The only concrete lacking was a solid block about 3 x 3 x 4 feet to be poured on top of one of the abutments.

About this time, along came Franz Forster, who was a construction man engaged in pouring concrete septic tanks and other concrete works around the new labor camps. Franz was on a motorcar which was hauling a car loaded with a concrete mixer. The mixer was churning away with a load of wet concrete. Franz explained that the concrete in the mixer was part of a batch left over from a nearby job which had been completed. To avoid wasting the concrete, he wanted to pour the block which was still lacking on the abutment. We examined the bridge plans together and decided that the block would in no way interfere with the erection of the steel. Norton, who was assistant to Bill Sprinkle on this type of construction, was overseeing the project of the temporary false work, but he was not about to offer an opinion about this, so we went ahead and poured the block.

The next day I was called aside by Bill Sprinkle who was Superintendent of Construction. He asked about the block, and I explained what had happened. He informed me that it was not customary to pour any concrete above the bottom of the steel work. He said that they would try to erect the steel with the block in place, but he doubted that it would be possible. He tried, but finally had to knock out the block before the steel could be installed. Bill was indulgent with me on this occasion, but, of course, he could not refrain from making some remark about my inexperienced engineers — the typical reaction of a crusty, old, practical construction man.

Shortly after this, Bill went on vacation. While he was away, we were in the process of erecting another smaller bridge crossing the Rio Siguacan right in the town of Tiquisate. By pure coincidence, Franz Forster came along with a loaded concrete mixer and asked permission to pour a retaining wall for the bridge. This time, the engineering office was handy, so we went over and consulted with Eddy Jones, who was the Division Engineer. We also consulted with Almer Bump, who was the Manager, and also an engineer. We all examined the plans and could see nothing to obstruct the steel erection. I reminded all concerned about Bill Sprinkle's admonition. Norton expressed no opinion. We went ahead and poured the wall.

Despite all the expert opinions, things didn't work out again. Bill came back from vacation just in time to see us tearing out the concrete. My explanation did not impress Mr. Sprinkle this time and he was somewhat less indulgent!

Boating in Almirante Bay

In 1953, Ed Spence was Assistant Manager in Almirante, Panama. Almirante is on a beautiful bay which is famous for good fishing and boating. One day while Ed and his wife were enjoying some boating, he somehow managed to fall overboard. Considering the fact that the bay swarms with sharks, the situation was a little hairy. Mrs. Spence had not learned anything about the operation of the boat and did not know how to stop it. All she could find to do was shut the valve to the gas tank. She then steered the boat in circles, keeping close to Ed until the gasoline in the carburetor was out. As soon as the motor stopped, Ed swam to the boat and climbed aboard. The next thing he did was to give Mrs. Spence instructions on the operation of the boat.

Another novel boating incident occurred to Raymond Stanton about 1954 in Almirante Bay. He was a Laboratory Technician in the Almirante hospital. Raymond, a French-Canadian, received his training in Rollins College in Winter Park, Florida. He was also a fit body builder.

Raymond was an enthusiastic boatman and had purchased a boat with a 25 H.P. outboard motor. Billy Graham (no relation to the preacher) had a small plywood boat which he had constructed himself with the aid of a kit. This boat was powered by a 5 H.P. outboard.

Billy got tired of his boat and decided to raffle it off. According to the old saying, "Them as has, gets," and as luck would have it, Raymond held the winning ticket. He experimented with his new toy by mounting his 25 H.P. motor on the plywood boat and cruising over to Bocas del Toro, which is on an island several miles from Almirante. On the way back, Raymond decided to open up the motor for a speed test. Much to his surprise, the light boat promptly disintegrated in the middle of the bay. By pure chance, the Company passenger boat, which operated daily between Almirante and Bocas, stopped to pick up Raymond and his passenger. After that, he didn't push his luck so hard and just bought lottery and bet on golf games.

Hacienda Tenorio

By

Carlos Colegial

Introduction

The following article appeared in a 1950 issue of the United Fruit Company magazine, Unifruitco:

"United Fruit divisions on the west and east coasts of Costa Rica were assured of a dependable supply of quality beef when, early this year, the Company purchased one of the choicest ranch sites in Middle America, in the highlands of western Costa Rica.

Named Tenorio (for a volcano at the ranch's northern border), the ranch has grazing land about 25 miles long that rises gradually from about 500 feet to 4,000 feet. None of the property is steep enough to cause erosion of the rich, black volcanic soil, and the upper two-thirds of the ranch enjoys excellent distribution of rainfall throughout the year, so that the pastures never dry up.

Thus far Superintendent John Steltzer has six cowhands to handle more than 2,000 head of cattle. As the latter figure grows to the anticipated 10 or 15,000, the personnel and facilities will expand proportionately."

This article featured photographs of the new ranch showing furniture arriving by air from the Quepos Division, the Steltzer's pet baby deer, mallard ducks, calves being branded, and the cool Steltzer home and headquarters building which was an old hacienda house built 1,200 feet above sea level. John Stelzer was shown paying ranch hands on the veranda of his house. The last photograph was of John and Mary Steltzer and Henry Hogaboom in a Jeep gazing out over the vast hacienda from the highlands.

Carlos Colegial Martinez was born in Rivas, Nicaragua, in 1914 and joined the pioneer group opening up the new Golfito, Costa Rica, banana division in 1939. When working in the Quepos Division in 1950, he married Maria Esther Arias Cruz. The young couple and baby Maria moved that year to the Tenorio ranch that don "Charlie" wrote about. As Division Inspector, he covered most of the departments on both coasts of Costa Rica where the Banana Company operated. After the banana operation in Golfito closed in 1985, don Carlos remained as the last Company employee in the port where he still lives in the house that he bought from the Compania Bananera.

Don "Charlie" Colegial was transferred from Quepos to Tenorio when the project was first started and provided the following first-hand account of Hacienda Tenorio.

Clyde Stephens

The Compañia Bananera de Costa Rica (CBCR) initiated Tenorio in 1949 from an old hacienda. The land covered 30,000 manzanas (21,000 hectares, or 52,000 acres) which extended from the hot lowlands to the cold, forested highlands on the western slopes of Volcan Tenorio. This property was located between Cañas and Bagaces in the northwest Province of Guanacaste and was bordered by the rivers Corobici, Tenorio and Chiquito. The Upala trail went through the property and crossed the Continental Divide to the remote Upala community near the Nicaraguan border. Much of the *finca* was covered in virgin forest and still had a lot of wildlife. Only a few patches had been cleared by small farmers for pastures, coffee and sugar cane. Basically, the area was still in a beautiful, wild state.

In typical United Fruit Company style, no efforts were spared in the development of this project. Since there were no roads, an airfield was made to fly in people and materials on the CBCR Company plane or on Lacsa Airlines. A squadron of carpenters was flown in to build houses, utility buildings and circular corrals, all made of fine local wood processed by the ranch's sawmill. Even the roofs were made of wooden shingles mass-produced on the property. All housing had running water, indoor toilets, baths and electricity generated by hydroelectric and diesel generators on the hacienda. Homes were stocked with some of the basic furnishings, which included leather-bottom beds. To supply provisions for ranch employees, rice and coffee processors were installed.

At one time, up to sixty employees, plus their families, enjoyed a good standard of living at Hacienda Tenorio. The Company provided a school and a well-stocked commissary which had milk, chickens, corn, rice, beans and all the basic necessities of life. A good supply of fresh vegetables was assured the year-round because the farm always had good irrigation water from the river. Workers received excellent wages and were the envy of surrounding neighbors.

Actually, the *Unifruitco* article did not reveal the original justification for this enormous ranch. After the end of World War II, the United Fruit Company geared-up to replant Gros Michel banana plantations in Central America again. New plantations required thousands of mules to haul millions of bunches of bananas — four bunches to the mule — from the harvest site to the railroad. In addition, mules were needed for overseers, timekeepers and foremen to ride in the farms during supervision. All total, about eighty mules were required on each farm. To supply this need, Hacienda Tenorio was purchased to mass-produce thousands of mules which would be shipped to Guatemala, Honduras, Panama and the Costa Rican divisions.

Amazingly, a sudden change in plans by the Engineering Department at the Company headquarters in Boston caused the mule production program to be cancelled. A corporate decision was made by the engineers to build a vast network of cableways in the plantations to transport harvested bunches of fruit on a roller device to the loading area. This mechanical system replaced animal transport of bananas and the mule breeding project was halted before any mules were ever born.

Now that mules were out and cableways were in, the Company had to change priorities to justify the existance of this huge spread. The only obvious alternative was to raise cattle at Hacienda Tenorio. In time, native creole cattle were sent live to the United Fruit Company divisions of Quepos and Golfito in Costa Rica to supply beef to the Company commissaries. This transfer meant long cattle drives on the trail to launches waiting in the Golfo de Nicoya near Taboga. These creole animals were replaced with 2,000 improved heifers and 180 Brahman bulls imported from Honduras and the United States.

To work these cattle, there were 300 mares at the ranch. Prize studs were brought in to improve horse breeding. These included a quarter horse from Honduras, a palomino from the United States, and a Peruvian stud.

To feed the cattle and horses, jaragua grass was planted on 2,500 hectares (6,200 acres) of the property. Pastures were enclosed with 65 kilometers of four strands of barbed wire fence. Cattle were dipped in vats every 14 days for tick and bot fly infestations until the pests were reduced to low levels.

In the hot arid lowlands, rattlesnakes were numerous. To reduce the population of these vipers, snakes were killed and the rattles were cut off and sold for one Colon each, or about twenty cents of a dollar. On some days, seventy rattlesnakes were killed on the ranch. The boss had cans with thousands of rattles that had been collected and paid for. Higher up on the cool slopes of Volcan Tenorio, there were other poisonous snakes, such as the *terciopelo, bocaraca,* and the *oropel.* Despite so many snakes, very few cattle were lost from snake bites.

Coyotes were also a problem because they killed young calves. A bounty of ten Colones, or about two dollars, was paid for each dead predator brought in. Sometimes, big cats came out from their homes in the vast tropical forest bordering the pastures and attacked cattle. For these predators, a bounty of 100 Colones (twenty dollars) was paid for a jaguar. A puma brought only fifty Colones, which was about ten dollars. Except for these predators, it was forbidden by the ranch superintendent, John Steltzer, of Arizona, to hunt any other form of wildlife on the property. To enforce this policy, there were tough guards patrolling the borders of Rio Tenorio and Rio Corobici. Consequently, deer were plentiful and even grazed among the cattle.

Every January, there was a four-day rodeo with bull and bronco riding, roping contests and other rodeo feats. Cowboys rode in for the well-known fiestas from far-away Liberia, Cañas, Bagaces and Tilaran. A couple of young steers were barbecued and big pots of fresh corn on the cob were boiled to feed the crowd. Visiting cowboys were housed in tents and were treated as good neighbors, even though some of them got a bit too wild at times.

Hacienda Tenorio lasted only five years. In 1954, this vast holding with some 7,000 head of cattle, plus hundreds of horses, was sold to Costa Rican buyers for a mere one million Colones, or about $200,000. It was obvious that the Company wanted to get out of the cattle business fast. Today, the land has been broken up into numerous parcels which have been slashed and burned. The magnificent forests, the jaguars, the pumas and all the other wildlife have long since disappeared. But even in this condition, the present value of those lands is worth untold millions. Fortunately, the few of us still left will always remember that unique experience at Hacienda Tenorio where we started to raise mules.

Carlos Colegial Martinez from Rivas, Nicaragua, and Heino Iter Chan, an engineer from Estonia, with a jaguar killed at Hacienda Tenorio, Costa Rica, in 1951. (Courtesy of Carlos Colegial Martinez)

A Connecticut *Yanqui*

By

Richard A. Washburn

After thirty-six years with United Fruit Company, I reflect back to some of my early experiences when I first came to the Tropics in June, 1950.

It all started when a favorite uncle, Otto Lippolt, a well driller living in Ridgebury, Connecticut, answered my inquiry in 1944 as to which foreign language I should study in high school. "Spanish," he said, "the language of the future." He then told me fascinating tales of his younger days when he had taken his percussion rigs to drill water wells for United Fruit Company's sugar operation near Banes, Cuba. The stories of how they moved their steel-wheeled machines between locations with twelve pairs of oxen, the *bueyeros'* (drivers') cries of *arriba la pata carajo cono,* and similar Spanish phrases that my uncle had memorized, sparked a latent desire in me to aim towards the Tropics. I continued studying Spanish along with my agriculture courses at the University of Connecticut. My summers at college were spent working on a farm in Cheshire, Connecticut, and the fact that I always ended up in the very top of a dusty hay mow only heightened my desire to seek my fortune elsewhere.

I applied for summer employment with UFCo in 1949, my junior year as an "aggie" student. The request was rejected. A letter from F. F. McManus of the Boston Personnel Department stated that students were no longer sent down for summer work, but that if I were still interested, I should apply for employment the following year. Hopes still high, I waited until March, 1950, to write again requesting a job upon graduation in June. Much to my disappointment, the reply was, "Nothing available at present. If something should develop later, we will let you know."

Richard A. Washburn was born in Connecticut in 1928 and graduated from the University of Connecticut in 1950 with a degree in agriculture. That same year, he went to the Tropics. After working with bananas in Honduras, he switched to the African oil palm operations which later took him to Quepos and San Jose, Costa Rica. Richard is an authority on stamps of Honduras and has won several awards for his presentations at international philatelic exhibitions. He has also published in philatelic magazines. In 1988, Richard and his Honduran wife, Merida, retired to Rhode Island.

By return mail, I explained that their previous letter of 1949 had implied that something would be available upon graduation and that I simply did not believe a company the size of United Fruit would be unable to place a person like myself who wanted to work for them.

Thinking the matter closed, I accepted a position as soil conservationist in the New Haven County Agriculture Extension Office. A couple of weeks later, however, I received a letter from the Personnel Department of UFCo in Boston stating that the Company would hire me as an apprentice overseer at the Tela Railroad Company in Honduras for $150 per month, contingent on passing the physical exam. Health was no problem, and so I returned the employment form with a photo. My employment was confirmed June 1, 1950 without ever having been interviewed.

A week later, I took the train to New York. Once at UFCo's Pier 3, I met Mr. Jasper Baker, a distinguished-looking gentleman, who introduced me to two other recently employed young men also going to Honduras. One was Furber Roberts, entomologist from Maine, who had been hired by the La Lima Research Department. The other was John Varick who would work in the livestock operation. Both Varick and I had recently registered under the Selective Service Act due to events in Korea.

We were given letters of introduction to the General Manager of the Tela Railroad Company, Mr. J. F. Aycock. We also received complimentary theater tickets to the Broadway play, *Mister Roberts,* starring Henry Fonda, after which we were to board the night pullman train out of Penn Station for Baltimore where next morning the S.S. *Musa* was scheduled to leave for Puerto Cortes, Honduras.

The sea voyage aboard the twelve-passenger steamship took six days. Fortunately, Miss Jennie Jewett was coming home for summer vacation to visit her parents, Dr. and Mrs. V. C. Dunlap. Dr. Dunlap was head of the UF Tropical Research Station in La Lima. I say fortunately because upon reaching Puerto Cortes it seemed that no one had ever heard of a Washburn coming to work for the Tela Railroad Company. The only way to get to La Lima was to tag along, so we boarded a steel wheeled vehicle (called a motorcar, no less) that ran on the rails. Of course, this vehicle had been sent to pick up Miss Jewett.

Puerto Cortes hadn't changed much from the exciting days of Guy "Machine Gun" Moloney and railroad engineer Lee Christmas, an American adventurer from Louisiana who ended up as Chief of Police in Tegucigalpa with the rank of *General de Brigada.* The main thoroughfare was the railroad track that ran from the banana loading wharf towards the Laguna trestle, then on to San Pedro Sula and La Lima. Along the shore were tin roof dwellings, mostly on stilts, which had an over-the-water walkway to the outhouse under which fat catfish swam expectantly. There were large wooden, two-story houses and ramshackle hotels, something right out of the early 1900's. The clacking of the motorcar's wheels (especially noisy if one wheel happened to be a bit out of the round), the wind in your face, exotic smells (some pleasant, some not), stopping at the call booths to request the right-of-way from the dispatcher, the *cruces* — side tracks — to let pass similar vehicles and steam engines pulling endless strings of banana box cars, all seemed so unreal, yet most exciting, to this young Connecticut *Yanqui.*

Upon arrival in La Lima, the hub of banana activity for the Cortes Division of the Tela Railroad Company, Roberts, Varick, and myself had lunch at Dr. Dunlap's house. Afterwards, I left to meet with the Assistant General Manager, Bill Van Diepen, who told me to forget about delivering my letter to the General Manager and explained that I was to work on the Progreso side of the Ulua River. He also expressed his doubts (via a cockeyed look) that I would be around for more than six months. I, of course, disagreed. From the La Lima office, a rambling wooden structure with overhead fans, I walked to the motorcar garage and boarded another motorcar that took me to Omonita where I was transferred to a ferry and crossed the Ulua River. The ferry used the water current via pulleys and cables to propel itself from bank to bank, a distance of some 200 yards. I was met on the Pregreso side by a *Señor* Midence and we drove to the main office of banana operations of the Tela Division in a vintage Ford station wagon.

The Progreso office was built up on cement posts. Stately royal palms lined the pathway to the steps. Here, I met Mr. F. J. G. Robinson, Chief Clerk and a former banana farm overseer in the 1930's. He introduced me around the office. Since it was then near four o'clock, he suggested we head for the Company clubhouse. On the way he explained what he thought one should know in order to adapt quickly to the new environment. I remember words such as *highballs, vales, rifa, trainmaster, cuchumbo, daimes, bufalos, cuatro reales, catrachos, planilla,* and *chingona,* all heard on my first day. I was promptly initiated into a chingona dice game with a group sitting around the table rolling for drinks and ended up paying for a round. To pay, I merely signed a *vale* — no problem with money; it would be deducted from my pay at the end of the month.

Speaking of money, I was surprised on my first payday a month later when I received silver coins, both Honduran Lempiras and U.S. half dollars, mixed in paper rolls. There were no bills in circulation in mid-1950.

The following week was spent becoming acquainted with how the banana business operated from the farm to local headquarters level and what all the reports were: rainfall, fruit, livestock, spray (Bordeaux mixture), acreage, quality control, etc. I was given a McClellan saddle, bridle, pads, and a newly-purchased pair of spurs (these were for my personal account). I had already bought a machete, had it sharpened and the point properly shaped to fit perfectly in a fancy leather sheath that hung on my belt.

Early one morning, I left Progreso with Don Carlos LeSage, the superintendent of the Urraco District who took me to Buenos Aires farm. Buenos Aires, Victoria, and Nana Chumba farms were located along a belt line and were thriving in those days. (Due to drainage problems, they did not last long.) I was to be acting timekeeper. A month or so later, they left me in charge of the farm office. With the able assistance of the overseer, Henry Murray, a Cuban, whose father worked for the Company's sugar division in Guaro, and spraymaster Fernando Servillon, we made the monthly *planilla* (payroll) which was six pages and three feet by three feet. This huge report was hand-written with ink pencil that smudged with the slightest drop of sweat. We added, multiplied, and divided by hand and used four carbon copies. This payroll was due on a specific closing date, but competition was such between farms that one tried to complete it, errorless, a day ahead of time to be the first farm to deliver. Woe be to the farm that could not balance their payroll and had to call for help from the headquarters office!

Life on a banana farm started early with a general phone call to all farms for the previous day's rainfall report. The farm foreman would have roll call and assign the 100 or so laborers to their daily tasks. We supervisors rode mules to go from one job to another. My job was to learn by actually doing all the things that related to the growing and harvesting of bananas. Pruning, circling, fertilizing, spraying, cutting, backing, and hauling the stems to the *bacadia* were all well programmed facets of the daily activities. The *raison d'etre* of our daily labors was the production of good-sized stems of bananas that were cut at a certain grade (diameter) and shipped to port in railroad cars. The aforementioned farm jobs that related to the end product were varied and required some physical as well as mental proficiency. Unless you have cut a large stem of bananas and lowered it gently to the backer's shoulder, you will never know the skill involved.

Diversion on the farms consisted mainly of the weekly 16mm movies shown to the farm laborers on an open-air screen. These films of the 1930's and 40's were either Mexican *ranchero* themes, or the old Hollywood westerns. If you remained in a location for more than two years, you would see the same films return again.

Another pastime was to become involved in soccer. Our Buenos Aires team, *Cuba,* eventually became quite famous and won the local championship.

I also enjoyed hunting iguanas and would explore the creeks surrounding the plantation in search of the large males, redish in color. You had to hit them square in the head, or they would dive off the tree branch and go under water. The skins were tanned in Progreso and made into elegant wallets or belts.

The *rifa* — a raffle — was a special kind of entertainment. On one occasion right after the paymaster had come through on the railroad pay car, Henry Murray and I went to a *rifa* at Victoria farm. To get there, we took the railroad track bicycle, a four-wheeled, pedal-powered vehicle with a board seat up front. We would sit on the front seat and the Carib yardboy would pedal. The *rifa* in Honduras was a gala affair consisting of a suitable prize — the best being a chrome-plated .38 special revolver. The owner would sell one hundred tickets, numbered from 00 to 99. These numbers would be spaced consecutively on a round, spoked, paper-covered, plywood wheel with a center hole that would be spun once on an upright post. A local official or other person of reputable esteem with the ability to shoot straight would stand off 20 paces and fire a bullet into the spinning spokes. The winning number was that having the major portion of the bullet hole.

There was a group of American and European apprentice overseers in the Tela Division's Progreso operation in 1950-51. Some of the names that come to mind are Jim Olivier, Harold Hawkie, Anders Olsen, Michel Ganty, Ray Hodgson, Peter

Scott, Dallas Johnston, Bobby Walker, Buddy Tucker, Ted Elsasser, Philip Ryder, Fred Heine and later, Jimmy Britt. We were spread out working on farms in the Guanchias of Progreso District, the Urraco District, and the remote Toloa District with Tigre Farm at the very tail end, some two to three hours by motorcar from Progreso.

All of us would try to get into town for the weekend. The rules were that one employee had to remain on duty out on the farm (unless there was a fruit cut and we all stayed), while the rest of us came into town by motorcar to play golf, go to the movies or a dance, or just have a few drinks and talk over the week's events. Our Honduran counterparts would be there too, and we made friends with our fellow clerks and timekeepers: Tomas Becerra, Victor Acosta, Tavico Pineda, Hector Cruz, Justo Galo, Raul Ugarte, Benigno Gonzalez, Virgilio Sanchez, Salvador Canales, and one of the Zelaya clan from Olancho.

Once in town, most of us would converge on the Progreso Club. The club had pleasant surroundings and was located next to the nine-hole golf course. The veranda bar of the clubhouse was beside the pool which featured a large balsa log, split in half, to float on or set drinks on. In those days, most of us carried revolvers or some sort of pistol. These were collected for safe keeping by the watchman stationed at the club entrance as ordered by a large sign in Spanish and English which read:

"*Dejen sus armas en el porton con el guarda.*"
"*Leave your arms at the door with the guard.*"

A few of the older overseers and district superintendents would leave their gun belt as directed, but it was known who still had a smaller firearm hidden under his shirt *por si acaso* — just in case.

Most of us were always welcome at Georgia Penn's apartment. Georgia, a well-respected and jovial gal, was the principal of the Progreso American School and she enjoyed entertaining the multinational group of trainees. Her favorite companion was Buddy Tucker, overseer of Rapalo Farm. Buddy was an affluent and cultured member of the family who owned Jacks Brewery in New Orleans. To pass the time in his lonely overseer house, a grand piano was shipped to him from the States and transferred to the Urraco District by rail car. Tucker had to remove the side of his overseer house in order to swing in the piano, an incredible feat if you've ever walked up the hill to that house in Rapalo.

The trick was to hoist the piano up from a leaning coconut palm, get the heavy instrument swinging and land it on the upstairs living room floor. This piece of engineering was done without incident. Afterwards, people in the isolated farm area were exposed to classical and operatic music performed into the night — quite a change from their traditional *rancheros*.

Weekends were, for the most part, congenial affairs, although sometimes they got a bit rowdy if we ended up in a "downtown" Progreso bar drinking Fundador cognac with coke chasers or Salva Vida beer. Occasionally, someone would try the local *guaro* that came in two colors: clear or tinted pale yellow. This usually meant difficulties later on keeping your buddy out of a fight, and seeing to it that he was at the motorcar garage in time for the jitney that took us all back to our farms on Sunday night.

On Monday morning, bright and early, some banana men were not too bright, but they had to be up and ready for that never-failing general telephone call to begin another week on the farm.

Editor's note: *Of the 15 foreign overseers mentioned, half of them — Washburn, Walker, Johnston, Elsasser, Britt, Heine, and Murray — married Honduran women. As of 1987, Washburn, Murray and Walker still worked for the United Fruit Company in the Tropics.*

Opening of the new Golf Club in Progreso, Honduras, April 6, 1935. This event was celebrated annually for the next 30 years until the club closed. Horseplay was mixed with sportsmanship at this "O.F." tournament — "O.F." stood for "Old Folks" in polite company. Otherwise, it meant "Old Farts." They are: (left to right) F. J. G. Robinson, Dudley Linton, Ed Cobb, Horace Prowse, Toller, George Bryant, Don Swofford, Pat Myers, and Joe Felix Aycock. Swofford and many others were brought to the Tropics as athletes but some later became banana executives. Baseball, golf and horse racing were highly competitive events among Company divisions throughout Central America. (Courtesy of Scotty and Millicent Crichton)

Progreso, Honduras, golf tournament, 1952. This mobile bar constantly made the rounds among the players. The ploy was to intoxicate the visiting team. Left to right: Dallas Johnston, Anders Olesen, Dorothy Geivers, Ines Cappel (right seat), Georgia Penn (driver), Philip Ryder, Buddy Tucker (sitting), Bobby Walker (pouring) and Mercedes Roth. (Courtesy of Richard Washburn)

Socializing in 1955 in the Ganty's Farm 16 overseer house in the Progreso District near Santa Rita, Honduras. Note the jaguar and crocodile hides, the wooden crank telephone, and the canvas roll-up curtains behind. Both ends of the living room were open for ventilation. From left to right: Ray Hodgson, Teddy Elsasser, Jeanette Ganty, Jimmy Britt, Georgia Penn, Lydia Britt, Teresa Hodgson, and Irene Prowse. Michael Ganty was mixing drinks. (Photo by Dallas Johnston)

Since the Progreso Club was far away by rail car, recreation was found in a homemade pool at the Farm 18 overseer's house near Santa Rita, Honduras, in 1953. The auditors never found an inventory number on this pool. Philip Ryder (left) sprays beer on Bobby Walker while Bobby holds a bottle of Salva Vida between his feet. (Courtesy of Dallas Johnston)

Research and managers meeting in Tela, Honduras, 1958. Many of these veteran banana officials began their careers in the Tropics in the 1920's. This group represents about 600 years of service to United Fruit. Left to right: (seated) Alfred Butler, Kenneth Block, Ed Y. Jones, John Silver, Lazarus Greenberg, George Munch, Harlo Von Wald, Norman Sanderson, Jessie Hobson, Robert Hildreth. (Standing) B. E. Bookout, Carlos Estrada, Bing Miller, Norwood Thornton, Jack Strange, Andy Holcomb, Elmer Bump, Clyde Delawder, Andy Hedges, John Kelley, Harry Stover, William Taillon, Vining C. Dunlap, Dave Cloward, Luis Sequeira, Arthur Hamilton. (Collection of Clyde Stephens)

The Rise and Decline of Lake Yojoa

By

R. H. Stover

All *bananeros* who come to Honduras eventually get to Lake Yojoa. Most come to love the lake in one way or another — for fishing, boating or just plain looking. I don't remember exactly when I first saw the lake after my arrival from Canada in June, 1951; it was probably the following year. But once I saw the lake, I knew it would always hold my attention as long as I lived. At 2,000 feet in the mountains, days are pleasant and nights are chilled. No one who loves the outdoors can forget Lake Yojoa.

For many years, the Tela Railroad Company had owned a large strip of land on the east shore. Well before 1952, the Company had constructed the old stone main dining room and kitchen. The property had one stone cabin, a pier, and wooden boats with motors. The Tropical Research Department also had a small camp on the lake as part of a forestry project.

It was an arduous trip to the lake in 1952. An all-season road did not exist so you put the Jeep on a flatcar pulled by a railroad motorcar to the start of the Tegucigalpa highway at Potrerillos. From there, the trip over a narrow, very winding dirt highway took an hour and a half to get to the lake. About a year later, an all-weather gravel road was opened from La Lima.

In the early 1950's, Lake Yojoa was pristine with pine forests covering most of the eastern slopes and broadleaf forests on the west side. There were two small villages: Pito Solo at the southern end and El Jaral on the northern extremity, both formerly linked by a ferry before the road was built as a war effort in 1943. One could paddle around the entire shoreline and seldom encounter anyone. The local Muscovy and brown tree ducks, as well as many species of migratory water fowl, were abundant. Crocodiles could be seen daily.

Robert Harry Stover was born in 1926 in Chatham, Ontario, Canada. After getting a Ph.D. from the University of Toronto, he and his wife, Ann, moved in 1951 to La Lima, Honduras, where Harry began work on Panama disease research for $350 a month. He is a world authority on bananas and has written two books on the subject. In 1985, Harry retired to his farm and laboratory near La Lima but still circles the globe as a consultant. Since 1970, he periodically retreats to the wilds of the Mosquitia lagoons where he has a cabin at Puerto Lempira in the eastern corner of the country.

The main attraction of the lake was the magnificient scenery, especially the ever-changing cloud formations over the forested Santa Barbara massif rising in the blue sky to a table-like plateau at 9,000 feet. Fishing for *dormilon* and *guapote* was also popular and it was no problem to catch large strings; the fish were very edible but not much as fighters.

Various biologists had explored the lake and suggested stocking it with a more sporting species of fish. One day, after a banana field trip in September 1953 in the Progreso Club, highballs and beer flowed and ideas became action. John Hollis deposited a bill on the table and suggested that everyone contribute to a fund to stock the lake with largemouth black bass from Florida. The bills continued to come, thanks to those present: George Poland, Pat Coon, Guy Cantrell, Lee Cox, Harlow Von Wald, Alphonse Chable, John Hollis, Bob Mixon and myself. Later, other employees, and especially Felix Aycock on behalf of the Tela Railroad Company, contributed generously to the largemouth black bass fund.

The story of the 1954-55 stocking and its huge success was told by Gene Ostmark in a 1964 San Pedro Sula newspaper story. I will outline some of the more noteworthy events and especially those that have led to the present state of the lake and its surroundings.

By mid-1957, some two years after the stocking of 1,800 fingerlings from Florida, occasional eight-pound bass were being caught and four-pound bass were common. Between 1957 and 1965 the bass population exploded and replaced most of the native fish. Lake Yojoa provided some of the most magnificient surface plug casting in the world. It was common to catch 20 to 40 fish per person in a day in the three to eight-pound range. Only at the peak of the hot, dry season in March and April did fish activity slow down.

While the lake provided hundreds of untold hours of glorious sport fishing for the *bananeros* and their friends for many years, much more significant events were impinging on the lake's ecology with far-reaching consequences. In the mid-1960's, the lake's natural outlets were closed to raise the water level. A canal was dug on the north end to take the additional water from the lake and channel it down the mountain in giant pipes to hydroelectric turbines at Cañaveral and Rio Lindo. Now, it was possible to partially control the water level and the large annual fluctuations of 15 to 20 feet of previous years no longer occurred.

Along with the hydroelectric project, a new highway, and a 3.5 percent annual population increase, came the people — thousands of them. New communities and small farms sprang up along all the shores during the 1970's. Numerous houses were built near the shoreline and some over the water. A long strip of shanty-like restaurants specializing in fried fish now stretches for 15 kilometers beside the highway overlooking the eastern shore of the lake. Most of the forests were cut down except for a small area of privately-owned broadleaf forest along three kilometers of the west shore.

It was not until 1974, about eight years after the peak sport fishing had passed for the *bananeros,* that foreigners and the United States bass-fishing tourists discovered Lake Yojoa. Fish Hawks International established a tour operation from New Orleans at the Hotel Agua Azul, which was formerly the old Tropical Research Department forestry camp. This was followed by another fishing tour company at Hotel Los Remos. The fishing tours generated an intense publicity campaign in U.S. sporting magazines and newspapers: "Home of the Next World Record Bass" *(Field and Stream);* "Found — a Lake That May Hold the Next World Record Bass" *(Fishing Facts);* "Anglers Hunt World Record Bass at Lake Yojoa" *(The Kansas City Star);* "Your Best Chance for a World Record Bass is Right Here" *(Dallas Metroplex Sportsman).* However, the best sport fishing had already passed. Bass larger than six pounds were seldom caught and a twelve-pounder was exceptional. The world record being sought was 22 pounds. By 1978 the bass fishermen's tours to Lake Yojoa had ended. An expensive new luxury hotel built to accommodate expected tourists was closed in 1981. From 1974 onward, bass fishing declined rapidly in both quantity and quality. Fish now served in the kitchens along the roadside are frequently salt-water species brought from the north coast of Honduras.

The reasons for the bass decline are not known. No studies have been made and no data on fish populations have been collected in ten years. The ecology of the lake has been changed by the hydroelectric installation and by the diversion of two streams into the lake in the late 1970's to add more water for the turbines. In 1964, tilapia were introduced and this fish of Asiatic origin is now plentiful. Some commercial netting of fish is done but it is unlikely that overfishing is a problem. It appears that some link in a biological food chain is now missing and the bass no longer grow to a good size. Only a biological study can define the problem and suggest a remedy.

Along with the decline of fishing, the problem of people, uncontrolled expansion of grazing, and lack of planning are everywhere. Lake Yojoa is no longer pristine and it is unlikely it ever will be again. The Food and Agricultural Organization of the United Nations along with government agencies prepared plans in 1975 and 1978 for rational long-term multiple uses of the lake. Proposals had been made in the 1960's and 1970's to establish a national park around the lake. The entire area was declared a Forest Preserve in the 1970's but this decree has been generally ignored. It would seem that as long as four different government agencies and numerous private land owners have jurisdiction over the lake and surrounding lands, nothing will be done except for preparing plans and projects that will never be executed.

For the thousands of us who have sat on the cabin porches mulling over Company affairs, enjoying the splendid view and the once magnificent fishing, it is sad to see what has happened. But it is enlightening. Lake Yojoa is a microcosm of what is happening around us: idiotic population growth, irrational land use and resource planning, and foreign aid policies that exacerbate these very problems that they are supposed to alleviate.

Although sad, I find it is still enjoyable, and at times exhilarating to look up at Santa Barbara or down into the schools of fish in the clear water. The beauty and tranquility is still there. I can recall the passing parade of thirty years, of friends who have gone, many forever, who were once my companions on these beautiful waters.

The Sula, a ferry of the Empresa Dean, docked at the north end of Lake Yojoa at El Jaral in 1930. This Lincoln car, which had traveled two days from the Honduras capitol of Tegucigalpa, continued on down the rugged mountain road to the railhead at Potrerillos and was loaded onto a flat car destined for San Pedro Sula. Ferry service terminated when a road was built along the eastern shore of the lake in 1943 as part of the war effort. (Courtesy of Larry Mahan)

The west side of Lake Yojoa was covered with rain forests which continued to the peak of Santa Barbara with an elevation of 2,744 meters. In contrast, pines dominated broadleaf species on the eastern shore where the Fruit Company cabins, boats and botanical plantings were located. (Photo by Margaret Hogaboom, 1947)

Lake Yojoa was teeming with fish and wildlife in 1947. Left to right: Mrs. Richardson, Jefe Richardson with a string of guapotes and dormilones, the motorcar man, Richardson's grandson with a big muscovy duck, and Henry Hogaboom holding pichichi wood ducks at the Company cabin. (Photo by Margaret Hogaboom)

The Flood

By

Iolene Butler

Tela Railroad Company
La Lima, Honduras
October 6, 1954

My Dear Children,

Last week we had a terrible flood here on the north coast of Honduras. You may have heard about it over the radio on the American news. I have had no opportunity to write you anything earlier than now.

It all started on Sunday, September 26, with a soaking rain all day. Nobody took much notice of it as this is the beginning of the rainy season and we have to expect a lot of water. The coolness was a welcome change as we have had very hot weather during September. The water had started to flood our yard, but that did not alarm us as we are situated in a low part of the Fruit Company housing zone in La Lima. Our house is about five feet off the ground and water has never bothered us too much before.

By the afternoon of the next day, we had plenty of water around the house and out on the golf course. I called the yard boy to help me move the chickens to a higher chicken house. To my annoyance, I found out he had already gone. He had heard reports that there was a big wave of water coming towards his house and rushed home to help his family. Without his help, I went to move the chickens myself but found that they were well above three feet of water. So I left them safe and dry.

When "Butt" came home about four, the water had risen. Before we had finished our tea, it was getting uncomfortably

---

Iolene Powell Butler was born in 1906 in England and married a British horticulturist, Alfred P. Butler. From 1928-30, they lived at Lancetilla, Honduras, where the United Fruit Company began developing the famous tropical botanical garden under Dr. Wilson Poponoe. After serving the Company in Jamaica, Iolene and Alfred transferred back to Honduras to another new UFC project and joined the original staff of the agricultural school at Zamorano from 1942-44. They also lived in Costa Rica and Panama. Since her husband's death in 1963, Mrs. Butler has lived on the Isle of Wight with her daughter, Wendy Birnie, occasionally crossing the Atlantic to visit her son, John, in Fort Myers, Florida.

high. Water had begun to wash the bottom step in the front. Butt looked at his car and said he thought he should move it to higher ground near the gate. I persuaded him not to, which, as it happened, was good advice. We looked out over the golf course and saw a big wave of water sweeping across the fairways and greens. In no time, the road between us and the golf course was awash. This also raised the water level under our house. It came so quickly that soon the cook in her *casita* out back was shouting to us to come help save the chickens. We took one look out back and saw it was sheer madness to try. We yelled at her to get back in her house and stay there.

By this time, the current was swift and was bringing in railway ties, oil drums and all kinds of debris. The next time we looked at the car, the water was half way up the side of the garage and the car was on its side. Later, the car rose with the water and had it not been under the zinc roof, it would have floated away like other cars did.

Since we were afraid that the water would rise high enough to get into the house, we started putting things up on the beds, unplugging lamps and taking clothes out of bottom drawers. Sure enough, water started bubbling up through the floor boards. Then we heard shouting out front. Some men driving home from the other side of the golf course had stalled their cars and had started to walk in what they thought was shallow water. When they got nearer to this side, they found they had to swim. One man got into difficulties but they got him safely into the house next door. Both men were strong swimmers but they said they'd never have made it had they not clung to two heavy railroad ties that came floating by.

We had just started rescuing of our things again when we heard some more shouting out front. Somebody yelled, "There's Butler's house." We still had electric lights so we turned on more lights to guide them in. They literally swam into the front door. One boy got caught in the current and was swept towards the garage. Fortunately, he hauled himself along the top of the hedge and barely made it to the door. We got them dry clothes and asked what on earth they were doing out in such dreadful weather. It appeared that they had had a frantic telephone call from this side saying a house was on fire. The husband had not gone home yet and his wife was scared. These unexpected guests started helping us once they took a straight shot of whiskey and warmed up a bit.

We put a table in the middle of the bathroom under the trap door in the ceiling. If the water got waist high, we would get up into the roof for safety. By this time, we had about given up saving anything, except for Pat's Cornell University diploma which he had worked so hard for. A bottle of Scotch was also salvaged for "iron rations."

The power had gone off and we just sat on the bed with our feet tucked up under us waiting for the water to rise. It was black dark but we had good flashlights. Occasionally, somebody would flash the light on the wall to watch the water level. By 10:30, it had stopped rising and by 11:00, the level receded a bit. That's when we decided to get some sleep. At midnight, the lights came back on again but we were scared to walk through the water for fear of getting electrocuted. The fuse box was under a few feet of water and all the wall plugs were under water. However, by six the next morning, almost all the water was gone except for two inches of oily mess all over the floors and walls. Also, there were grimy rings up on the furniture. What a dreadful mess!

That morning, we had trouble getting the maid and her little girl from their *casita.* The water was too high for wading. In fact, it was then about eight feet deep in the yard. They had no food and, of course, were scared to death. We finally got a couple of fellows with a boat to go and fetch them about ten o'clock. Soon our guests departed by swimming out to shallower water and then waded up to their waists.

The next day, Butt's foreman sent over six men to help clean out the house. I don't know how much soap and disinfectant I used. The men came daily to clean up and we kept them filled up with plenty of hot coffee. By Saturday, we had the place fairly presentable. But the smell is still terrible and I can't imagine what it will be like when there is just a slimmy pool left.

Naturally, we felt sorry for ourselves, but people on the line out in the farms had it tough. Some people sat on roof tops almost a week. Many also lost all of their children and possessions. President Eisenhower declared it a distressed area and sent an aircraft carrier with helicopters into Puerto Cortes. The helicopters have kept up a steady service, dropping food where they can, taking doctors out and bringing sick and injured people in. Everybody is being inoculated against typhoid. Where people cannot get in, they are flying nurses out to them. It is funny how we are so accustomed to seeing helicopters landing on the golf course and at the hospital. Most people had never seen these machines before.

A week later, we had more rain and the yard began to flood again. We went to bed at nine but were awakened about midnight by our neighbours who were usually very quiet. They were running back and forth through their house and doing a lot of shouting. It was hard to hear what they were saying because the frogs were making so much noise. Soon we got a telephone call from them reporting that the *comandancia* was warning of a bigger flood on the way. As tired as we were, we started putting things up high again. This time I wanted the maid and her daughter in the house with us and started calling her. The neighbours were distressed by my shouts and sent their hired man over to help me. I was on my own because Butt had driven the car we were now using to a higher spot. Then the yard boy came and helped get the maid into our house. Once again, things were put up higher. Finally, we got some sleep.

A bit later, there was another warning of a new hurricane but fortunately flooding was not bad in our area. Although we heard thunder rumbling again, it was a heartening sight to see helicopters at the end of our golf course just in case. Connie Platt called and said Ernest phoned and told her to start putting things back up high again. He heard the rivers were rising again.

The cause of all of this trouble was Hurricane Gilda. The storm was expected to hit Belize but it turned around and caused a cloud burst in the Quimistan Valley. Our rivers were already high, but this tremendous volume of water came down with so much force that the levees broke in several places. One little village a few miles up the valley was completely wiped out, so they say.

Like I said, we were luckier than many, but we did have some losses. We lost all 33 laying hens and four hives of bees which Butt had just rehabilitated. Our furniture and a lot of other stuff in the house is ruined or badly damaged. Practically everything in the garden is dead or dying.

People on the street in back of us in W-type houses are bad off. The water stood almost to the ceiling and several cabinets containing dishes were tipped over. Some refrigerators were ruined. The Kirkconnels had bought Tom Tidy's piano and had hoisted it up on chairs but to no avail. It tipped over and was a total loss. Some houses, like the McNair's, got a horrible deposit of grease and oil. All those with deep-freezers under their houses lost everything in their freezers.

I thought the Banacks would get a lot of water and invited them to stay with us. Surprisingly, they stayed quite dry on their side of the golf course near the higher bank of the Chamelecon River. The Petries and Aycock houses stayed dry but the Van Diepens got four feet of water under their house. Even the Thorntons had it to their floor boards because the levee broke by the Turnbull house and at Guaruma.

Naturally, a lot of things went floating around. Mattias found a full five gallon can of gasoline in the corn patch. It must have gone right over that high wire fence. The Smiths from the construction department found their washing machine out in the bananas. Dr. Kingsbury's car floated from under Dr. Lynch's house and over the high fence into the yard next door. I am mighty thankful that no dead body floated in. That has happened to some.

Many of the banana farms were heavily damaged. This, coming right after a nine-week-long labor strike this year, is going to be a bad year for the Company here. They say the Tela side has not been flooded so much and three of the higher farms remained above the water level.

Many of the farm people have suffered terribly. There are stories of whole families lost. In La Junta, people have been on their roofs for a week now. There have been some rescue operations by boat from Puerto Cortes with the Company launches — the *Sula*, the *Chamelecon* and the *Rover*, collecting people and issuing food.

During the last run of one of the Pagan buses from La Lima to San Pedro Sula, 27 people were drowned when the bus turned over. Some of the survivors slept in trees during the night.

We saw some pictures taken by Mr. Auler of the Rio Chamelecon at flood stage. Before the crest, the water was already up to the iron bridge at Lima Vieja. By that time, houses and people had already been lost.

I forget to tell you about Paco the parrot in all this. When Cruz, the maid, went out last Monday night, she found that the parrot had fallen into the water. Cruz rescued him and took him into her room. When the boat went to fetch them, Paco was left behind. He was quiet until the afternoon when he started call ing us. First, he called "mummy," which can be for Cruz or me, but when that got no results, he started calling "Mrs. Butler!" Cruz managed to get out to him on Wednesday morning, feed him and put him back in his tree again. On Friday, Daddy noticed that Paco had moved to a lower branch of the avocado tree but he wouldn't talk. He was wet and a sorry-looking specimen. Soon, the six o'clock whistle blew and that perked Paco up a bit. As soon as Cruz came out and slammed her door, Paco started talking. She sloshed through all that water and mud and carried him back to his tree, both of them talking baby talk all the way. After that, he stayed up there and wouldn't come down again.

The La Lima Company hospital is simply full and crowded and everybody is working at top speed. Everyone is being vaccinated against typhoid. Fortunately, we had our booster last August. Yesterday, one of the crew aboard the aircraft carrier was operated on aboard the ship for a ruptured appendix. He was in bad condition and was flown by helicopter to our hospital here in La Lima.

Well, that's all for now. I'll send you more news later. Don't worry about us because we are safe and fine. Let's hope the worst is over.

Love,

Mum

We Lived on a Banana Plantation

By

James O. Harrison

It was in 1956 that the big white ship on which we were traveling made its way slowly toward the dock. Overhead the thunder rolled and rain poured down. We were approaching the port of Golfito, Costa Rica, where we would soon leave the ship and go to our new home. My family and I were going to live on a banana plantation.

I am a biologist, and my new job would be to help control the insects that destroy the plantation's crop. Spraying with deadly chemicals often kills helpful insects as well as those that damage bananas. It was up to me to find ways to control insect pests without using so many chemicals.

When we were ashore, we were surprised to learn that there was no automobile road to the plantation. We would be taken by railroad. Of course we expected to ride on a train. But then we saw our baggage being loaded into an open station wagon.

My eight-year-old son Michael asked, "How can we ride in a car if there is no road?"

We soon saw that this car had no rubber tires. Instead, it had iron wheels like a train and it ran on railroad tracks.

Soon we were on our way, listening to the clatter of the wheels as we rode along. By this time the rain had stopped and the sun was shining. For a while we traveled by the edge of a dense forest, where we saw colorful birds and many beautiful butterflies. Also, lizards seemed to be everywhere. Soon, however, the tall trees gave way to a forest of banana plants.

Our car slowed to a stop. As the driver got out to switch the track, he said, "We will be side-tracked here for a while until the banana train passes."

James Ostelle Harrison was born in 1920 and grew up on a farm near Harrison, Georgia. In 1956, he arrived with his wife and son in Golfito, Costa Rica, on the *Morazan* to work on banana insect problems at the Coto 47 Research Station. Later, Jim went to Cornell and returned with a Ph.D. in entomology. As the "father of banana ecology," his research led to the practice of natural insect control without the use of insecticides in banana plantations. In 1962, Jim joined the staff of Mercer University at Macon, Georgia, and has broadened his interest in natural history through teaching and travel. He retired in 1987 but continues to explore the world with his wife, Joyce.

Soon we heard the train coming and as it passed we counted almost forty freight cars loaded with bananas. They were on their way to Golfito to be loaded on the ship that had brought us to Costa Rica.

We finally arrived at the plantation, where we were met by a large group of men, women, and children. One boy had a pet spider monkey sitting on his shoulder.

When we reached the house that was to be our home for the next few years, we found that it was built on posts about eight feet above the ground. During the long rainy season everything on the ground stays so wet that there is a serious problem with mold and mildew.

As we went to the house, a large iguana scampered across the grass. Michael spotted another one up in a tree. We were soon to learn that these large lizards are abundant on the banana farms as well as around the settlement. Iguanas can grow to be more than three feet long. They eat only leaves and other vegetation.

The large plantation where we lived was made up of twenty farms, each with its own settlement. All of the farms were connected by roads, but there were very few automobiles. The roads were mainly for hauling bananas. When people wanted to go someplace, they usually walked or rode bicycles. Sometimes a bus ran between farms and the plantation headquarters.

We lived at the headquarters where a Company store, a small hospital, a church, and two schools were located. At one school all of the classes were taught in Spanish and at the other all classes were in English. Michael was certainly glad to know that he would not have to learn a new language before he could start school again.

The workday on a banana farm begins at 5:30 a.m. and there are many jobs to do. There are drainage canals to be cleaned, plants to be fertilized, young banana shoots to be pruned, underbrush to be cleared away, insects to be sprayed and fruit to be harvested. The workday usually ends about 1:00 p.m. because of the heavy afternoon rains.

When a ship is on its way to pick up a load of bananas, it sends a radio message. So, long before it arrives in port, the farmers know it is coming, and begin cutting the fruit that is ready to harvest.

No ripe fruit is harvested, however. Any bananas left on the plant until they were ripe would split open and spoil during the trip to market. So bananas have to be harvested while still green.

Since each banana plant produces only one stem of fruit during its life, the plant is cut down when the fruit is harvested. While one man cuts the plant, another catches the bunch of bananas on his shoulder as the plant falls. The fruit is then carried to the nearest road and placed on a flat trailer. When the trailer is full, a tractor pulls it to the packing station.

At most packing stations today the fruit is washed by sending it through a small building very much like an automatic car wash. Then clusters of bananas are cut away from the main stalk and packed in boxes. From the packing station, the fruit is carried by train to the seaport and loaded aboard a refrigerated ship. Sometimes it takes all night to load one of the big freighters. But within twenty-four hours after the bananas are harvested, they are on a ship headed for market.

Published in *Highlights for Children* in January, 1980.

A banana ship from the Great White Fleet leaving Golfito, Costa Rica. (Collection of Clyde Stephens)

James O. Harrison beside a granite sphere in the Palmar, Costa Rica, banana plantations in 1960. The largest ball recorded was 2.44 meters in diameter. The origin of these monoliths remains a mystery. (Courtesy of James Harrison)

Digging Into Costa Rica's Past

By

James O. Harrison

A bright flash of lightening crackled through the air overhead and brought me back to reality. I was hot, dripping with sweat, caked with dirt, and at the bottom of a hole about five feet deep. I had not noticed the approaching storm because for some time my attention had been fixed on a polychrome vessel at the bottom of the hole. My mind had wandered back through time, hundreds of years into the past, as I slowly freed the vessel from its resting place.

This was my first attempt to excavate an ancient Indian burial in Costa Rica and with a beautiful vessel in sight, I wasn't about to stop because of a little rain. However, the downpour soon became more than a little, so I climbed out of the hole and let the cool rain wash the dirt from my arms and legs. After a few minutes the shower passed, and I went back into the hole to continue digging. By the time darkness forced me to stop, I had recovered four complete vessels and many pieces of a rather large pot which I hoped to reconstruct.

This was in 1956 when I accepted a position as research biologist with the Compañia Bananera de Costa Rica. I hadn't dreamed that I would be exposed to the prehistory of southwestern Costa Rica so soon. But on my third day in the country, while being shown through banana farms near the Rio Coto, I saw several places where someone had been digging and the ground was strewn with hundrds of pottery sherds.

I asked my guide, Mr. Paul Hari of the Banana Company's Research Department, about the holes and pottery sherds. He told me that some of the natives had been digging Indian graves.

As Mr. Hari began to tell me about the many graves that had been dug in and around the banana farms, I could not conceal my excitement. But, the more he told me, the more alarmed and dismayed I became at the tremendous waste and destruction of valuable archaeological materials. Local natives were doing all of the digging, and they were mostly searching for gold artifacts. Beautiful pottery vessels were often broken open to reveal their contents and then discarded.

In order to establish a new banana farm, all of the trees in the area must be cut down. Banana seed pieces are planted and soon afterwards, workers throughout the farm begin digging the many drainage ditches that will carry away excess water during the coming rainy season.

Much of the land in Costa Rica that is now covered with banana plants was once cultivated by pre-Columbian Indians. They lived in these low flat lands, cultivated their crops, and buried their dead. Even when the Spanish *conquistadors* came to this part of the New World, Indians were still living and fighting their wars on this land. But since the coming of the Spaniards, Indians have disappeared from many areas and tropical forests have again covered the land they cultivated.

When laborers on the banana farms dig drainage ditches, they sometimes dig through an ancient burial ground. The word spreads. Other workers rush to the site, and they soon dig up the entire area. Sometimes, but very infrequently, a gold artifact is uncovered. This is enough, however, to keep the natives fired up in their search for treasure.

While Mr. Hari continued to drive me around the banana farms, he told about one burial site where only a few graves had been dug. He stopped the jeep and as I started to get out, he said, "This is Farm 44 and the place I was telling you about is just across that creek to the east of here. You can find it very easily by just following that trail until you come to a place where some people have been digging."

I learned later that the area east of Farm 44 was not suitable for bananas because of a hard layer of limestone about three feet below the surface. *Señor* Jose Varela, the foreman of my field crew, also told me how to locate a grave in that area. He gave me a long iron rod that was sharpened at one end with a short iron bar welded across the other end to give a T-shaped handle. When the iron bar is pushed into the ground, it will only go down about three feet, to the hard limestone layer, unless it is being pushed into a grave. In the latter case, the rod will go easily a little more than four feet into the ground. It seems that the original diggers of all graves in this area cut down into the limestone at least one foot.

I was able to locate and excavate three graves in this limestone area which is about fourteen miles west of the Costa Rica-Panama border where it is crossed by the Interamerican Highway. Each grave contained four or five pottery vessels, one or more effigy whistles, and one or more stone celts. There was no sign of bone in any of the graves.

Most of the pottery vessels were decorated with red and

black lines on a white background. This color pattern is typical of alligator ware from the Chiriqui province in Panama. Three small tripod vessels were of a uniform brown color. Two of these tripod vessels had solid legs, but the legs of one were hollow. Each hollow leg had a narrow longitudinal slit through which a small round clay ball could be seen. The balls were free to roll back and forth in the hollow legs as the vessel was tilted.

Three of the effigy whistles were bird-shaped and one was in the form of a dog.

One of the graves contained a clay figurine that is typical of a common Chiriqui style. It appears to be a seated woman with legs widespread and hands resting on thighs. This figurine is also decorated with black and red lines.

Several months later, Jose Varela and I were exploring the headwaters of the Rio Corredor, just a few miles northeast of the Coto banana district. While walking along a narrow ridge overlooking the stream, we came to a series of excavated graves, right on top of the ridge. The graves seemed to be arranged in a definite pattern and each hole was surrounded by many sherds. We skirted open graves for more than a quarter of a mile, and I have wished many times since that I had counted them.

As we studied the pattern of opened graves, there seemed to be one place where there should be another but the site had not been excavated. We decided to gamble and spend a few hours digging at that spot. Most of the opened graves seemed to be about six or seven feet deep, so we knew it would take several hours to get down that deep.

The ground was relatively soft and the digging was easy except for tree roots which happened to be in the way. At a depth of about six feet, we found a leg bone so we knew that this was definitely a grave. We continued to dig with a hand-carved wooden trowel and found several more bone fragments. Then we came to the first vessel, a plain brown pot with a globular body and cylindrical neck.

After recovering two more similar pots, we struck a piece of brown porous stone which turned out to be a stone statue. The statue is twelve inches tall and appears to represent a priest with a sacrificial knife in the right hand and a victim's head under the left arm.

Then, while removing dirt from around a piece of skull bone, I saw a glint of yellow. I slowly pulled the object from the loose dirt and could hardly believe my eyes when I saw in my hand a beautiful gold bird-shaped object. While still removing dirt from around the piece of bone, another glint of yellow proved to be a thin gold disk about three inches in diameter. When weighed at a later date, the disk weighed 5.45 grams and the bird weighed 11.77 grams.

It was very interesting to note differences between the artifacts from this grave and those from the graves near Farm 44. The two cultures were undoubtedly separated more in time than in distance. Even though bone fragments were found in this mountain grave, it was probably older than the graves I excavated near the banana farm. Apparently soil conditions have more to do with bone deterioration than age of the grave.

Sixty miles northwest of Coto lies the banana district of Palmar. Here banana farms are located on the extensive flood plain of the Rio Grande de Terraba — also known as the Rio Diquis. When workers from the Banana Company cleared the forest from this land, they encountered many stone balls ranging in diameter from a few inches to more than seven feet. The balls were arranged in definite patterns and probably numbered more than a thousand. We will never know the exact number or exact pattern, however, because many of the smaller balls have been carried away and some of the larger ones have been blasted apart by people looking for gold.

No one knows who made these spheres nor how they were made nor for what purpose. One thing is certain, however, most of them were made by master craftsmen. Some of the larger balls are almost perfect spheres and are believed to weigh as much as sixteen tons.

On several visits to Palmar (the banana district headquarters is located at Palmar Sur, on the south bank of the river), I was invited by some of the workers to go digging with them. On one occasion, while looking around some old diggings on an abandoned farm, we were lucky enough to locate an undug grave. From this site, we recovered three shallow tripod bowls of polychrome ware and one taller tripod pot of brown ware. We also recovered a small jaguar-effigy metate that was just ten inches long overall. Its small size, and the fact that it shows no sign of wear, seem to indicate that it was constructed strictly as a ceremonial object. This grave also contained several other vessels (shallow tripod bowl types) that were crushed into many pieces.

While I was at Palmar, *Señor* Mario Volio told me many interesting things about the Banana Company's early days on the Duquis Delta. He told me that when workers first discovered the stone balls, some of them were located singly, but most of the spheres were in groups, the largest of which contained forty-five balls. In some places, the balls rested on small platforms paved with river stone.

Most of the balls were made from granite but at least three are known to have been made from a very hard type of coquina. It is still an unsolved mystery as to how these balls were transported from their place of origin. The coquina balls are at least twenty miles up river from the nearest raised beaches where coquina can be found. And the granite balls must have come from the mountain foothills across the river. Only small granite boulders, washed down by the river, are found anywhere near the present site of the stone balls. It is believed that rafts were used to float the balls upstream or across the river.

Señor Volio also told me about many small stone statues that were found in the forest when it was cleared away for banana farms. Most of the statues were made from a kind of white limestone and were arranged in an upright position around the perimeters of low mounds. Most of the statues were between seven and twenty-four inches tall.

The low mounds proved to be foundations for dwellings. It is, therefore, very possible that the stone balls which were located on platforms were actually resting on the foundations (floors) of long removed dwellings. These balls may have been kept inside the dwellings.

For those who visit Costa Rica and have only a short time to be there, The National Museum in San Jose has an excellent collection of artifacts from the Diquis Delta and Coto regions. One of the stone balls is displayed in the courtyard of the Museum.

For those people who have their own transportation, the Interamerican Highway crosses the Rio Grande de Terraba at Palmar Sur. Banana Company roads connect with the highway there and lead to many of the large stone balls which still remain in their original locations.

Published in *Popular Archaeology* in January/February, 1975.

Oklahomans in the Tropics

By

LaVerne Caid

Since our arrival in the Tropics with the United Fruit Company in 1957, many memorable things have happened. I always recall the tense times before, during, and after the Trujillo assassination in the Dominican Republic in 1961. Stressful personal stories include when I saved our yard boy's life by rushing him to the hospital after finding him lying in a pool of blood. He had attempted suicide because his wife ran away with another man. Also at death's door was our maid who had attempted an abortion on herself. She, too, was lying in a pool of blood when I found her, but recovered after being rushed to the hospital.

We lived in La Lima, Honduras, longer than in any other place. My husband, Russ, traveled for the Research Extension group for many years and left me alone with three small babies. Like most people in La Lima, I decided it would be safer to have some armed protection around the house. Consequently, a pistol was bought during our next vacation in Oklahoma and smuggled back to Honduras.

Ever so often, there was a shootout between political or military factions and bullets went flying everywhere. Soon after I got my pistol, a notice was circulated that a house-to-house search for guns and ammunition was being made. I was scared to death that I would be caught and had decided to bury my pistol and bullets in the back yard. Much to my relief, the police called off the search before reaching our house.

During these tense times, the chickens caused another scary experience. Russ was gone, as usual. For some reason, I forgot to lock the chickens up in the chicken house. Since it was dark, I didn't want to go out by myself so I just left them alone. About eleven that night, I was just getting to sleep when the flock began squawking nervously. By the way they sounded, I suspected that opossums were out for a meal. All I could visualize

---

LaVerne Caid was born and raised in Oklahoma and came to La Lima, Honduras, in 1957 with her entomologist husband, Russell D. Caid. The Caids also lived in: Puerto Libertador, Dominican Republic; Tiquisate, Guatemala; Golfito, Costa Rica; and Changuinola and Puerto Armuelles, Panama. Besides raising three sons, LaVerne was active in community projects. They left Honduras in 1987 when Russ accepted an executive position with Chiquita Brands in Cincinnati.

was those bad animals taking bites out of my poor chickens and couldn't stand it any longer. I got up, went out the back door with my pistol and headed for the chicken house. Sure enough, my flashlight shined the eyes of opossums trying to eat the chickens. Taking reckless aim, I fired several shots into the darkness of the chicken house thinking that surely I had killed an opossum or at least scared them away. After hurrying back to the house, it wasn't long before the chickens starting squawking again. I must not have done a very good job after all. With all of this noise in the night, my neighbors became alarmed and rang the Company telephone as soon as I got back into the house. They were relieved to find out that I was doing the shooting rather than being the one shot at.

Before we bought the pistol, Russ bought a single-shot air rifle to keep in the house. One Sunday night after he left on a business trip, our large watch dog started barking and running around the yard which was well-lit. The noise kept our two-year-old Steve awake so I lay down beside him. As Steve and I lay quietly in the dark, we began to hear funny noises in the back yard. Without turning on the lights in the house, I looked out the window to see where the dog was, but there was no dog. Instead, about eight tall, black men went streaking across the back yard . . . completely naked with not a stitch on. All I could think of was that they had killed the dog and now intended to come in for me. I had to act fast.

For some reason, I had loaded the air rifle that night and laid it on a shelf in Steve's room. Normally, the gun stayed unloaded. Trembling, I took the rifle and walked to the window, and yelled in the loudest and bravest voice I could muster, "Get out of here or I'm going to kill you." Then I fired the single shot which sent them scurrying behind the servant's quarters. Terrible things went through my mind. I was afraid they would realize that I didn't have a real gun and would come back again. When trying to reload the gun, my hands were shaking so badly that the pellets fell all over the floor. Finally, I fumbled around long enough to get a shot into the chamber. As it turned out, I succeeded in bluffing those naked intruders away. Apparently, they had swum across the Rio Chamelecon from their side at Lima Vieja, which explained the reason why they had no clothes on. After I caught them, they swam back to their territory again. I suspect that the mango tree in our yard was partly responsible for some intrusions during the fruiting season.

But there also were many happy memories too. In Honduras, fishing with the family at Lake Yojoa or in the sea at Tela was always fun. All three boys, Steve, Bruce, and Neal, had their own fishing tackle and knew how to catch fish. I also think of our pleasant vacation trips back and forth from the USA on banana ships. In those days, we had six weeks of vacation: one week up on the ship, four weeks in the States, and one week back to the Tropics aboard a ship. In the Dominican Republic, we have memories of eating hot toasted cashew nuts while driving across the countryside. We also remember the Christmas trees without leaves because we lived in a desert on the north coast of the island where barren trees were cut for the occasion. Also in this desert terrain was the Grenada Company golf course in Puerto Libertador (named in honor of Trujillo — "the liberator"). The greens on this golf course were groomed sand and that made putting more interesting.

Teacher in the Tropics

By

Phyllis Stephens

"I'm going to teach in Honduras next year." My family, needless to say, was aghast. Did I know what I was doing? And where was this place? I must say that I had to look for Honduras on a map when I was offered the job.

After teaching several years in Minnesota, I decided to travel. I went to Port Angles, Washington where I taught the second grade for one year. I loved it there but still wanted to travel. My friends in Glenwood, Minnesota, where I'd taught the year before, sent the school newsletter. In it, I found an ad for unusual teaching opportunities. One of the companies listed was the United Fruit Company. I applied for a position and in a few weeks I was called and offered a position in Progreso, Honduras, teaching grades seven and eight. I accepted, and then told my family in Minnesota.

I got what information I could — which wasn't much — about the place where I would live. The brochure I was sent was out of date. People did not wear suits and long dresses anymore. Luckily, I decided to forgo these.

In August 1961, my adventure began. I took a train from Minneapolis to New Orleans and then flew to Central America because there was no ship available at the time. Fortunately, a young man from the banana divisions was traveling on the plane and he was told to help me with connections in Guatemala. Without him, I would never have been able to change planes for Honduras. I landed on the old grass airfield in San Pedro Sula and was met by the principal, Marie Chalk. She informed me that I would be teaching grades six, seven, and eight.

I was given a two-bedroom apartment in Progreso close to the Gulf Club. At the door of the club, a sign read, "Leave your arms at the door with the guard," in English and Spanish. Many

---

Phyllis Irene Steffen Stephens was born in Bellingham, Minnesota, in 1932. After becoming a teacher, she taught eight grades in a rural one-teacher, one-room school. She arrived in Progreso, Honduras, in 1961 to teach in the Tela Railroad Company banana school. In 1963, Phyllis married Clyde S. Stephens, a research entomologist. She has been the principal of Fruit Company schools in Changuinola and Puerto Armuelles, Panama, and in Golfito, Costa Rica. Her hobbies are cooking, arts and crafts, sewing, and reading. She is presently teaching in La Lima, Honduras.

of the members carried pistols and the rule was enforced. The school was also very close, about one block from the apartment. What a difference in life style! For twenty dollars a month, I got an English-speaking maid who did everything — cooked, cleaned and washed. After doing all of this for myself for several years, I enjoyed the luxury.

I still remember my first day at the school. I was not used to the names. Like many girls in Honduras, one of my students was named Suyapa, after the patron saint of the country. I don't remember how I pronounced it, but the students practically rolled in the aisles laughing. I also had a boy named Adalberto. I just could not get my tongue to say it right, so he became Bert.

It was an interesting two years. I had taught all eight grades by myself in a country school in Minnesota so teaching multiple grades in Progreso didn't bother me. I did find that my students had many local beliefs. A black butterfly in a house meant death. One child told me it was so, as one got into their house and their mother fell downstairs. I heard numerous stories about the man without a head. A girl told me that her father had seen him and even talked to him. At night there was no way they would walk across the golf course alone. There were several graves in the middle of a fairway and the ghosts from these graves walked at night. One girl even insisted that she saw a woman in red lying on the grave in the middle of the day. The woman in red disappeared when she got near. The girl avoided that area even in the day. Then there were the little people who appeared in houses, generally in the bathrooms. The children were deathly afraid of them but apparently the little people usually caused no harm. When I said that I didn't believe in them, they told me something bad would happen to me!

I started to play golf in Progreso. I hit very long balls, but my direction was terrible. So was my game when I got to the greens. Several people offered to help me. I was told that I should not lose my temper when I hit a ball incorrectly. Later, I saw these same people throw their clubs in anger when they hit a bad shot. Actually, the lessons didn't help much; in fact, I got worse!

One strange thing about the golf course was the small cemetery in the middle of it. One had to shoot over it or be out of bounds if you landed near the graves. This was the same cemetery where the children saw the ghosts.

I enjoyed teaching there, but decided that I needed to move on. I resigned and intended to work on my degree and then perhaps go to Europe. One day soon after this, I had to see a doctor in the La Lima hospital. Getting there was an adventure. There was no bridge over the Ulua River so we had to cross the river on a ferry, or if the river was high, in a *cayuco* — dugout.

While I was in La Lima, I met E. J. Wehunt, who had been on the ship with his family when I returned to Honduras the second year. E. J. and Lorna were having a party that night, and asked if I would come. I wasn't really prepared for a party but went anyway. I met a young man from research. He was with someone else and she implied that they were serious. I thought, well another interesting one taken. To my surprise, he called me within a week and asked me to a *despedida* — farewell party. I accepted and when we got there, I found out it was his *despedida!* He had been transferred to Changuinola, Panama.

We had a whirlwind courtship and were soon engaged but decided not to announce our plans yet. Clyde left, we corresponded and my personal belongings were sent to Panama. I wondered what the people were thinking. This was before it was common to live together and no one knew we were engaged at the time.

Clyde came back to Honduras in July, 1963, and we were married. But before the wedding, there were some problems with the marriage application. After filing, we waited and waited. Finally, the private Progreso lawyer explained the reason for the delay. He said that we could get married, but since we were related, it would cost more money. This excuse was puzzling but we finally caught on to what was happening. The Honduran lawyer and bureau officials were phonetically pronouncing our names in Spanish, but were not looking at the letters. Orally, Steffen and Stephens do sound alike but Clyde pointed out the difference in the spelling of the two names. They agreed that we were not related and we soon received the papers just before the wedding. We were married "legally" under the law of Honduras by the *alcalde* — mayor — of Progreso in the Departamento de Yoro. This civil ceremony was witnessed by Libby and Leland Davis. Unfortunately, Clyde doesn't remember very much because of his bachelor's party the night before. His well-wishing gang of men friends brought him home at sunrise and we were married at nine. That farewell-to-bachelorhood party was talked about for years.

The next day, we were married in the little Union Church facing the golf course in La Lima. The last wedding there had been seven years before and there was a big turnout of friends. Bill Newell gave me away, Shirley Rarden was the matron of honor and Bill Rarden was the best man. Our Swiss friend, Paul Hari, sang a love song in French. Afterwards, there was a lovely reception in the home of Russ and LaVerne Caid.

The honeymoon was in the Bay Islands. To get there, we drove on the new gravel road to Tela, then flew on Sahsa the next day to La Ceiba and boarded a small passenger and cargo launch to Utila. Our first night on the island, we almost slept in separate rooms because Mrs. Rose's Guest House was full. The women were in one room and the men in another. At the last moment, it was discovered that we had just been married and a Peace Corp girl kindly moved to a man's room and he wound up sleeping on a couch somewhere else. At least we had a three-quarter-size bed out on the porch, but the rate was the same — less than two dollars per day per person, including breakfast. The islanders, British Anglo descendents, spoke a funny local English and held a grudge against England for ceding them to Honduras a century ago.

We also had some interesting times on the larger island of Roatan. There was a man who also was just married, but by force. He spent most of his time roaming around outside in a fatalistic way, and he even serenaded us at night. One night we went into a very small shop for a drink. I sat on a chair partially hidden behind the door. Clyde sat on the counter looking out the door. We were talking to the owners when a black woman, known as "the brown bomber," came in, slapped her hands on my husband's thighs and said, "How about a drink, baby?" The owners hissed and kept pointing to me. She finally looked around, saw me, laughed, and left. We saw her several more times around town and when she passed us, she always hummed, "I'm in the Mood for Love!"

In Guanaja, also known as Banacca, we went to a dance one night. Our isolated motel was reached only by boat and our group hired a boat to go to town in the dark. We all had a good time at the dance hall. The next morning, Clyde and I decided to go back and see the town in the daylight. We were surprised that the sea was full of rocks and reefs where we'd gone so fast in the dark the night before! That day when we arrived, the boatman said he needed money for gas and that he'd meet us in about two hours. We waited and waited. No boatman. Clyde asked a woman where he lived. When she heard the story, she said that he was a "liquor-head" and that by now he would be drunk. We found his brother, the mayor, who apologized and volunteered to take us back to the motel in his sailboat at no charge. Sailing back was much more fun than the motorboat.

From 1963 to 1973, Clyde and I lived in Changuinola, Panama. Our two children, Linda and Carl, were born in the banana port of Almirante. During this period, we also found some interesting folk beliefs. I had a maid who informed me that the reason that Linda had a virus was because we had too many people over and that one had given her "the evil eye." The only way to counteract this was for me to urinate in a cup and give it to her to drink. Needless to say, I refused this form of folk medicine.

We had another maid, Guillermina, with a bizarre background. The peasant girl was from back in the jungle and had no hair. She lost it when she was about seven years old. The local witch doctor had persuaded her parents that their daughter was possessed by a demon. In order to purge and purify the poor girl, they dressed her in a white robe and scarf while she was growing up. It didn't help and she grew up hairless. When Guillermina worked for me, she was about 17. One day she went to the hospital with me to help with the children. Dr. Nicola Messina saw her and asked me about her. I told him the story. He then asked to see her head. He prescribed an iodine solution as he thought she had a fungus. Sure enough, she was cured in a few months and her hair grew out again. She proudly stopped wearing the scarf and was extremely happy with her new head of hair.

There was another superstition that was common in several countries. If you swim on Good Friday, you will either turn into a fish or get eaten by them. In Honduras, the children used to stand around the pool on Good Friday to see us turn into fish. Our children loved the water and were snorkeling at age two. I was very surprised one Easter vacation when Carl, then five, would not go into the water. Before, you couldn't keep him out. Finally, he said that his friends in kindergarten had told him that if he swam on Good Friday, the fish would eat him. I finally persuaded him that this would not happen and he reluctantly came into the water. When we'd had enough snorkling for the day, I said to him, "See, the fish didn't eat you."

He replied, "No, but there were a lot more fish out today than usual!"

During our time in Changuinola, Panama, we always had *cayucos,* hollowed-out log boats. We frequently went to the San San or Changuinola rivers. I remember one of the first times we were out in the San San River. Clyde, Ruben Castillo, and I were cruising around when we spotted a troop of monkeys. To get a closer look, we took off for the bank. Clyde wanted me to jump ashore, grab the rope and pull the boat up. He yelled, "Jump," which I did, and landed knee deep in mud. We did get a good look at the monkeys, but that was the last time I obeyed the captain's orders to jump.

In 1973, we moved to Golfito, Costa Rica. What a wealth of superstition! Our children insisted that there was the *Llorona*, a woman whose child had died and she was still looking for it. She hated red and my children refused to wear the color. They told me that she had come and sat on the hood of a car that belonged to their friend's mother. The friend's mother beat the *Llorona* off with an umbrella. At night the children would get together and tell ghost stories; then everyone was afraid to walk home. They would walk each other home but the last one was terrified. One night I took a group of children out to observe stars. They had a good time and were very excited until we saw a bright light on the horizon. I think it was Venus, but they insisted that it was a spacecraft. Even the boys stuck close to me and I had to walk each one home.

One afternoon Linda celebrated her birthday. That evening they were to go to a bonfire and Clyde and I were going out to a dinner party. We were just about to leave, when all the children showed up at the house because the bonfire had been rained out. Could they listen to music? Since we were only going across the street to the home of Martha Jane and Tom Honiball, we said yes. Several hours later, I got a frantic call from my daughter. Vera Flores couldn't see. I rushed back home and found Vera sitting in a trance, staring straight ahead and repeating, "It touched me, I can't see." Obviously, she was hysterical. After questioning the others, I found that they had been telling ghost stories, then decided to play "hide and seek." Either she ran into a spider web, or more likely, some older boys touched her with something soft in the dark. She knew it was a ghost and became so frightened that she couldn't see. I took her and all the smaller ones home and told the rest they could stay, but NO MORE STORIES! The kids were happy to comply as they were all scared to death by then.

Today, we talk and laugh about all the superstitious things that nearly got those kids. Our children, both grown now, laugh too but say it was real to them then.

We began our married life in Changuinola, Panama, where we lived for ten years. After nine years in Golfito, Costa Rica, we spent nearly six years in Puerto Armuelles, Panama. Now we have gone full circle and are back where it all began — in Honduras. It has been an interesting twenty-eight years in Central America.

The dormant Bocas Division was revived when Manila hemp (abaca) was planted during World War II to make rope. This photo shows the superintendent's house in the small, isolated community at Farm 8, Changuinola, Panama, in 1943. The Clyde Stephens family lived in this house in 1971 after the farms were replanted with bananas. The area surged in growth when headquarters was moved from Almirante to Farm 8 in the 1970's. (Collection of Clyde Stephens)

The 1969 research and managers meeting at the old Golf Club in La Lima, Honduras. This assembly represents many disciplines in the banana business and portrays a rich culture of some 16 nationalities. Meetings were in English but nearly everyone spoke Spanish. (Collection of Clyde Stephens)

Jon Roberts wearing a Chiricana hat and Marty and Bill Bosy at the fair in David, Panama, in 1962. (Courtesy of Mary Bosy)

Water from melted snow irrigates bananas near Santa Marta, Colombia, the only place in the Caribbean where snow-capped peaks are seen. These coastal mountains tower up to 19,000 feet. (Photo by Scotty Crichton)

The Avocado

By

Billy and Kathy Saravanja

When most United Fruit Company kids finished the eighth grade in tropical American schools, going on to high school and college meant boarding a banana ship or plane and heading for one of the southern states in the U.S.A. There, eight years of our lives were spent among people who had no idea where our homes were nor what kind of life styles we grew up with. The year 1970 found us at the University of Southern Mississippi at Hattiesburg with our educational career nearly completed.

Kevin, a fine lad from Tupelo, Mississippi, and wife Donna, a Southern Belle, lived in the on-campus married quarters. Being friends of ours, they decided to invite us to a home-cooked dinner. The date was set and we eagerly looked forward to the occasion as one gets tired of typical university cafeteria food. (We were single at the time.)

Prior to the evening of the dinner, while Donna was preparing the meal, she decided that it would be nice to include a Central American dish. As we sat down at the table where everything was immaculately laid out, there was no question about the fact that Donna was a fabulous cook. She explained her idea about the special tropical dish she had prepared along with the meal and asked us to taste it and identify it. It was served in a small bowl and resembled some sort of dip with an off-tan color.

After several samples, we could not recognize what this Latin dish was and began to think that perhaps Donna was pulling some sort of joke by preparing a Southern recipe and telling us it was a Latin dish. But I ruled out this possibility as we had spent our high-school and most of our college years in Mississippi which exposed us to every Southern dish from vittles to grits.

Finally, we gave up tasting and asked Donna to tell us what she made this dip from. I think she was a little indignant that we were unable to identify one of our native foods. She went back into the kitchen and came out with a hard, green avocado in her hand and said she was saving this to make another dip. Still puzzled over all this, we asked her how she had prepared the dip, to which she replied: "First, I peeled off the outer skin with a knife, exposing a light yellow pulp which I tasted, but found to be hard and tastless. So I continued to shave the green part away with a knife until I came to this large, round seed in the center which I boiled for two hours. Then I removed the seed and discovered it was still hard, so I cut it up in chunks and boiled that for one more hour. The seed chunks were still hard so I ran them through the blender, took them out, mixed them with mayonnaise, mustard, Lea and Perrins sauce, diced onion, and — presto! — "avocado dip."

We both looked at each other and burst into laughter. We then explained to Donna the facts about avocados. The seed is what gets thrown into the garbage can, not the pulp, to which she replied in her Southern drawl, "Oh, my goodness. Y'all have to forgive me this time." We were, however, invited back to eat at Kevin and Donn's again when the second avocado turned ripe.

Impressions of a Spray Pilot

By

Tam Syme

I arrived in Changuinola, Panama, on March 27, 1969, after training as a banana spray pilot for two months on the Pacific side at Puerto Armuelles. My first impressions of the place were not at all good, especially since the pilot's house in Farm Eight was in poor shape. Upon driving to the house for the first time, I was aghast to see plastic fertilizer bags over the windows, but I learned that the bags kept in the air-conditioned air. Fortunately, my wife and two children were not with me or I'm sure that they would have turned around and gone straight back to our new home in London. After living with the Royal Air Force for twenty years, this was not the kind of housing we were accustomed to.

Within two months, the house was renovated, thanks to the help of the Chiriqui Land Company. Nevertheless, when my wife arrived, she nearly refused to stay, but I convinced her that our future was here with ATOPAN, a contract company which sprayed bananas for the United Fruit Company. I soon adapted quite well and became a member of the "banana man" team. The two children also adjusted quite readily. In fact, my son, Ian, was tops in his Spanish class of Latin Americans, thanks to his teacher, Doña Eva Pena. When my daughter, Ann, went to Saint Andrews University in Scotland, she studied Spanish and French. Thus, my children did well with languages.

My wife, Frankie, had a bit of difficulty learning Spanish. However, she was suddenly forced to learn more of the new language when the Medical Superintendent, Dr. Nicola Messina, put her in charge of the Farm Eight medical clinic which had been recently installed in the old Changuinola Golf Club building. This job was a complete culture shock for Frankie who had been an officer in the Royal Air Force as a registered nurse.

Tam Syme was born in Edinburgh, Scotland, in 1928 and became a jet pilot in the Royal Air Force. Most of his 20 years in the service was spent in the Middle East. He met another officer, nurse Audry Frankland, in Cyprus during the 1956 campaign against Egypt. A year later, they were married in England. He was a liaison officer for the Ministry of Defense in the development of the vertical takeoff Harrier jet. As a spray pilot, Tam has flown cotton in the Sudan, potatoes in England and bananas in Panama where he became a citizen. The couple retired in Spain in 1986.

As everyone knows in learning a new language, there are certain "slips of the tongue" that make the face turn red sometimes. This happened to Frankie when a woman came to the clinic to have a baby. Many of the young Indian mothers were very young and she always asked their age. She said they always seemed quite surprised when she asked each time. One day, she was assisting another nurse and doctor with a delivery. She noted that the woman was very young and asked her age, as usual. The doctor began laughing helplessly. Frankie asked what was so funny. He told the nurse, "You tell her," between bursts of laughter. It seems that she was pronouncing an *n* rather than an *ñ*. Instead of asking, "*Cuantos años tiene usted?*" (How many years do you have?), she was asking, "*Cuantos anos tiene usted?*" (How many anuses do you have?). After one patient being questioned answered, "*Solo tengo un ano, señora!,*" Frankie replied that surely she had more than one. When the Panamanian nurse explained the mistake in pronunciation, there was some embarrassment, but that was all part of the learning process.

Frankie performed a valuable service in the Farm Eight clinic until the new government hospital in Changuinola replaced the clinic and the old Fruit Company hospital in the banana port of Almirante.

Spraying some 17,000 acres of bananas is rather like painting the Golden Gate Bridge — you start at one end, and when you finish, you start all over again. Consequently, life for the spray pilot can be hours of boredom spraying flat plantations of bananas. However, boredom can sometimes be punctuated by sudden moments of absolute terror. During my twenty years in the Royal Air Force flying single engine jet fighters, I'd experienced some near-misses — being shot at by Egyptians in the 1956 Six-Day War over the Suez Canal. Also, I was shot at by rebel tribesmen in the Aden Protectorate. As the leader of an RAF acrobatic foursome, there were many thrills but these were executed with precision and discipline. However, as a spray pilot, I have truly been scared more often than I care to remember.

Our spray planes were Snows, named after the designer, Leyland Snow. The aircraft I flew were powered by Pratt and Whitney 600 HP Wasp engines and these planes glided like a brick when there was no power. The engines were mostly built in the 1940's during World War II and had been overhauled and sand blasted so many times that you could nearly see through the cylinder walls.

My first "heart-in-the-throat" experience came in 1971 while flying over bananas along the Sixaola River. The engine of HP 381 suddenly stopped and white smoke billowed from the front. As taught in RAF, I instinctively thought of the emergency checkpoints for a forced landing but there was no time to apply any of the emergency procedures. As the plane was quickly settling down into the canopy of banana leaves, there was a sudden jerk — the tail wheel had caught onto a banana cableway and we came to an abrupt stop. It all happened so fast that I didn't have time to get scared until later. I quickly got out of the cockpit unharmed, but the plane was a mess, as was my underwear. After checking the cockpit, I realized that I had instinctively cut off the mags and fuel cock, an effort to survive. Only a few dozen banana plants had been destroyed as the plane stopped suddenly, thanks to the steel cable. Bobby Walker was General Manager then and didn't seem to mind the loss of a few plants. An investigation revealed that a bearing in the supercharger had failed. Even though this was not a pilot error, my dignity had been shattered.

This wrecked plane was hauled out of the bananas and was flying again after many months of overhaul. Unbelievably, I crashed down into the bananas again, two years later, in that same plane when the engine stopped on me for the second time. The cause: a bad bearing in the supercharger again! Fortunately, I walked away unscathed, but it occurred to me that the finger of fate was punishing me for all of my evil ways — past, present, and future. Tony Pease, the new General Manager, examined the accident and was pleased that only a few banana plants had been lost. The plane was once again pulled out and overhauled. We have both mellowed a bit with age and the both of us are still flying. There were other frightening prangs, but too many to write about.

Flying had its dangerous moments, but there were good times to be had with so many different people in Changuinola. Don Harris, born and raised in the Tropics by British parents, was chief engineer. Jim Stacy, an Old Guard banana man, was the head of agriculture. Eric Price, British, married to a Panamanian lady, was assistant manager. Harry Brown, ex-dragline operator, was in charge of heavy equipment. The machine shop was run by Ray Murphy, descendant of a British-Hindu cross, born and raised in India. His wife was a Moslem from Malaysia. Fitzgerald was an Irish railroad man who married a Chilean. Chris Millensted, British, was a seasoned banana man and now was manager of production. The research department was run by Clyde Stephens. Coming up through the ranks were Gregorio "Chicho" Gomez, Raul Romero, Roberto Barnett, Godofredo Miranda, Ruben Castillo, Abdiel "Chicharon" de la Lastra, Guillermo Guerra, Cornelio Guerra, Miguel Chock, and Leslie Brown. They would eventually replace the foreigners and become the first generation of Panamanian directors and managers in the Bocas Division.

Some of my first experiences in Panama were with Clyde Stephens. I enjoyed the fine tropical cooking by his wife, Phyllis, and learned about the local terrain, fauna, flora, and native inhabitants from Clyde. Every object and subject had its own wonderful story to tell. I've never known anyone who immersed himself quite so deeply in his surroundings. Consequently, everything we did together was like an adventure.

The Changuinola golf course was quite a challenge but it was a relief getting to hole "19" — the club bar. Here, a staggering amount of Scotch was consumed and the bar bills were shocking at the end of the month. That is, unless you were lucky at rolling chingona dice for drinks.

During 1970, there were four severe floods when the rivers overflowed into the plantations and housing zones. The Base Line air field was barely above water so I could fly out with emergency food and supplies in the spray tank. These materials were dropped to the hundreds of victims stranded for days on the roofs of their houses. Except for those lucky people who lived high up in stilt houses, hundreds of people lost everything they owned. It was a horrible time but, fortunately, only a few dozen people were drowned. After the flood waters receded, it was an awful mess. The only plus was that the fresh silt deposits were beneficial to the plantations later on.

While I was not flying supplies to victims during the April flood, I rather had fun cruising around the streets in Finca Ocho, the "Snob Hill" housing zone, in my motor boat ferrying people around and running errands. At each house I pulled up to, Paul Johnis, my engineering friend, and I were invited in for a drink. You can imagine that after so many ports of call, this led to some rather ridiculous behaviour. By the time I motored back to my own home, I was feeling no pain. But I was strongly rebuked by my poor family who was sitting on the house steps clad in life vests waiting to be rescued themselves. The water had come up into our house about seven feet but luckily we lived in a high stilt house. Since Chris and Hildegard Millensted lived in one of the new single-story chalets, I evacuated their family to our place by boat where they spent the night in safety.

The floods were devastating to the banana operations: thousands of acres of bananas were lost, offices, machinery and engineering shops were left in a mess and hundreds of houses were devastated. Cleaning up and repairing the damage cost the Company a lot of money and effort.

Boating in the rivers, bays and shores of Bocas del Toro was wonderful. There were memorable picnics at Bocas del Drago on Isla Colon, water skiing at Knapps Hole ("bare-ass beach") and three-day excursions with Clyde Stephens and Sammy Brent in Clyde's dugout. On one occasion, the three of us planned months in advance to make a seven-hour trip out to a lonely island called Escudo de Veraguas. Clyde was the weather expert and calculated that the calmest time of the year to go out to sea would be the first part of October. Once we started, there was no turning back. But maybe we should have. The sea was wild with giant swells, chop, wind, and heavy rain. I was in the front of the long dugout with my airplane compass guiding Clyde, who was in the back steering his 25 HP motor and bailing out water as fast as possible. The climax of the trip was reaching the calmer leeward side of the island.

We slept, ate, drank, and fished in the rain for two days. Amazingly, there was a Guaymi Indian turtle fishing in a tiny *cayuco* — dugout — on the turbulent windward side. As we passed by his small boat, we cynically asked him how the weather was. His answer in pidgin English: "Not so baad today, baahs." That put us in our place and we kept on fishing in the huge swells, chop and rain. Despite the weather, we managed to catch some nice fish. The return to Almirante Bay was awful but the weather calmed down by the time we reached the capital of Bocas de Toro. All of this just proves that the weather in Bocas is unpredictable. You might say the same thing about spray planes.

Notes From a Banana Breeder

By

Phil Rowe

Why, Anyone Could Fix That!

There is safety in numbers — especially in the number of mechanics taken on a boat trip to the keys some 20 miles removed from the comforts of the mainland of Honduras.

The trip to the keys off Belize in Earl Guidry's boat in 1975 should have been guaranteed against minor mechanical failures because the passenger list was a "who's who" of La Lima's mechanical whizzes: Earl and his assistant were both helicopter mechanics, Lee Darlington had been a mechanic in the army, and Laban Hobson could fix anything related to electronics. I was a banana breeder and an amateur mechanic.

So, who do you suppose fixed the starter when it would not function for the trip back to the mainland? The lighthouse keeper on the key fixed it! These mechanics will not soon forget the advantages of growing up without ready access to a spare parts store around the corner. Just leave it to an isolated lighthouse keeper.

Touch Football a la Crunch

One of the great things about the Tropics is the international atmosphere created by the employees from around the world. This diversity in cultural backgrounds caused some confusion, however, in a touch football game in La Lima in 1970. Henry van der Wiel, a Dutchman, was a defensive back on the first play of the game. Rick Williams, a Canadian, ran out for a long pass. All six feet and six inches of Henry proceeded to dismember the 130-pound Rick. Henry had a look of satisfaction that comes from a job well done and Rick was reduced to a mass of writhing

---

Phillip Rowe was born in 1939 and grew up in his native Arkansas. After getting a doctorate in plant breeding at Michigan State, he and his wife, Jeanette, joined the Tropical Research Department in La Lima, Honduras, in 1969. Soon after, two sons were born in the Tela Railroad Company hospital. Phil has dedicated himself to breeding a perfect banana and, consequently, is a world authority on banana genetics. He also tends to the famous collection of hundreds of banana types acquired from all over the world. Phil continues his program under a new organization, La Fundacion Hondurena de Investigacion Agricola. After hours, he plays golf and chingona.

pain. There was a temporary lull to explain the difference between touch football versus tackle football in the real game that Henry knew only from movies. Rick was changed from wide receiver to a more sedentary position and the game went on by touching rather than crunching.

Flashback Memories

Our son quit wetting the bed because the maid told him she would put him in the oven if he ever did it again.

We made four specific trips to San Pedro Sula to get some film out of the *aduana.* In the end, the customs duty was only two centavos — one cent of a dollar.

While my wife was waiting to cross the old La Lima bridge, a drunk climbed into the car and started hugging her. He was cheered on by onlookers. Finally, the bridge was clear to cross and Jeanette escaped rather shook-up.

Forty Days Around the World

By
Timothy Johnson

Research Department
Tela Railroad Company
La Lima, Honduras
January 30, 1978

Dear Clyde,

This letter is about a jet-age saga of banana globe-trotters. To begin with, back in November of 1977, I was peacefully going about my business in Honduras when my reverie was shattered by a call from the New York office. It seems they were having some problems with shipments of bananas from the Philippines to the Middle East. Some buyers were complaining about high levels of unmarketable fruit and making large claims against United Fruit. My job was to rush over to the Philippines to find the cause of the trouble and then travel on to the Middle East to see if a problem really existed.

I wondered if that was the logical order to approach the problem, but, in any case, I flew from Honduras to San Francisco, Guam, Manila and thence to Davao City on the southern Philippine island of Mindanao. Needless to say, one suffers from a severe case of jet lag after some twenty hours in the air. Nevertheless, I busied myself sleuthing out the cause of the yet unestablished and undefined problem.

There was a welcome party given for me but actually it turned out to be a hail and farewell party. On my third day there, the New York office advised me to rush to the Persian Gulf immediately to inspect a load of bananas arriving in Dubai. Lloyd Berg, in charge of research in Mindanao, was to accompany me. Since time was short, my friends prepared a crash course for me in the gastronomic delights of the Philippines.

Timothy James Johnson was born in 1939 in Woodford, England, and studied in London, Trinidad and Michigan State University where he got his Ph.D. in horticulture. His wife, Eleanor, a Canadian anthropologist, also received her doctorate at Michigan State. Before working with banana research in La Lima, Honduras, in 1975, Timothy held posts in Nigeria and Kenya. The family now lives in Canada.

I had always been intrigued by tales of the durian fruit in the Far East. The natives believe in the aphrodisiacal powers of the durian but it is probably most famous for its smell and flavour. The trick amongst its enthusiastic fans is to forget about the incredible stink while enjoying the delicious sweet pulp which is sucked off the seeds. I was treated to this tropical delight but was not told how to separate the incredible stench from the delicious taste. Nevertheless, the economic botanist in me was satisfied.

The worst was yet to come. A colleague and friend, Dan Funk, formerly of Honduras Research and now working with our Company in the Philippines, the Mindanao Fruit Company, was on his way to the party with another special gastronomic delight known as *balut.* As the party progressed and tongues loosened, I learned about *balut.* This food is a specially pickled duck egg which has been fertilized and contains a well-developed embryo inside the shell. In the streets, one may buy a *balut* in the younger stages after fifteen days of growth. Or, if you prefer, an older embryo of twenty days, just before hatching, is also available. Obviously, the duckling has fuzzy plumage and a normal head by that time. It is said that the beak is kept to pick the teeth with after the delicacy has been eaten. Maybe it was the power of my prayer, but Dan Funk didn't make it to the party with the *balut* for me to try because his truck went into a ditch on the way over. Guess that delight will have to wait until the next trip to the Philippines.

Lloyd Berg and I flew to Dubai from Manila via Hong Kong, Bangkok and Delhi, finally arriving in the Persian Gulf, or Arabian Gulf to some. We based ourselves in a hotel in Abadan, Iran, at the head of the gulf because United Fruit has an agent at nearby Khorramshahr. This is just across the river from Iraq. There is a lot of wealth in these ports considering that Abadan has the world's largest oil refinery. Here, we met up with an old Honduras hand, Marvin Davis, who is looking after the Fruit Company's interest in the region.

Although foreigners have been warned to speak with discretion when discussing sensitive matters, Marvin didn't seem to mind making open and frank descriptions about "Eye-rainians" and the country in general. One of his most obvious traits is his stentorian Deep South voice which seemed loudest in the most public places. Lloyd and I were quite nervous about this and fully expected some of the Shah's strong men to come knocking on our hotel door at any time. But true to his character, this didn't bother Marvin at all.

Our intended mission was not working out as planned. In fact, we had not yet seen a banana ship with our fruit. However, there was a ship with fruit belonging to one of our competitors. The captain was very hesitant but finally allowed us aboard after we convinced him of our general concern for the international banana trade.

After inspecting some boxes of our fruit off-loaded much earlier into a buyer's store, we became quite frustrated because of poor communications and planning. However, there were things to keep us entertained. Once a film crew and cast spent some time at our hotel. We gaped at Curt Jurgens and Peter Graves of Mission Impossible fame. I think Marvin Davis even lent some tennis balls to the daughter of one of these stars.

After so many days of waiting for our banana ships, bore dom set in, but this condition was broken by frequent abortive trips to other Persian Gulf ports. Once, we were on our way to Dubai via Shiraz, Iran, but missed our connection in the latter city. After some resistance, the airline finally agreed to give us accommodations until our flight the next day. They explained that there were no suitable hotels in the city and that we would have to sleep on the outskirts of town. This made us very suspicious and a bit apprehensive.

At dusk, Lloyd and I, along with other stranded passengers, were loaded into a Volkswagen van and were driven to the outskirts of Shiraz. We continued for miles into the desert, seeing nothing but pure sand with an occasional palm tree. Just as we began to worry about our plight, the van pulled into a long palm-lined driveway leading to a huge hotel. It turned out to be the Inter-Continental Hotel near the ancient city of Persepolis, home of Persian kings Cyrus, Darius and Xerxes. Since we had the morning free the next day, several hours were spent exploring the ancient ruins that Alexander the Great had wrecked in 480 BC.

There was another memorable experience at this hotel. It seems that some of the staff, waiters and maitre'd must have been recruited from surrounding Muslim villages. The fact was clearly demonstrated during one of our meals when one of our party at the table was missing a knife at his place. This was no problem for our host — he simply went over to a nearby table recently vacated and brought over a freshly-used knife. In plain view, the waiter deftly wiped the knife between thumb and forefinger before delivering it to our table, good as clean. I tried to remember if he cleaned it with the right or the left hand.

During our five weeks of wild goose chases around several gulf ports, we found ourselves in the oil-rich desert island of Bahrain. Fortunately, we were given a nice tour around the island by an English lady and reviewed some more ancient history.

Christmas was getting nearer and I didn't fancy myself staying in the area any longer. We had not accomplished our original mission, and, in fact, we didn't inspect a single banana ship with our fruit. I got permission from the New York office and returned to Honduras, thus making a complete trip around the world. I don't know what happened to the undefined banana problem but no doubt it resolved itself.

Hope you had a nice Christmas.

Your truly,

Tim Johnson

The Promoter

By

Tim Moxon

I first met Bernie on a fine sunny morning outside the clubhouse at the Upton Country Club, Ocho Rios, Jamaica. I parked the little red sports car and hefted my bag of clubs out of the boot. The tall black boy was sitting on a low stone wall peeling a mango and he gave me a warm grin as I shouldered the clubs.

"Yu wan caddie sah? Carry yu clubs fe one dollar sah." He laconically swung his legs off the wall and ambled over. The caddie looked about twenty-three, and the way he walked reminded me of the tiger in the ads for Esso gasoline, all sinew and poised and mildly predatory.

"Sure thing," I replied and handed him the bag. "Never seen you around here before. What's your name?"

"Bernard, sah, but dem call me Bernie. Mostly work ober at Runway Bay Club but dat a tourist place and ain't nah tourist dis time of di yeah."

Bob and Fred and Jonesy were already in the locker room when I went to change. All three had that palsied look of a banana spray pilot on a Sunday morning after the festivities of a Jamaican Saturday night. I greeted them gaily and changed into my golf shoes, black and white brogues — very natty but they didn't improve my game one damn bit.

The other budding Arnold Palmers had golf carts and when we crunched out over the gravel driveway, Bernie was waiting on the first tee cleaning the head of my driver with a piece of rag. It was already warm and his ebony face shone with a mild sweat. Bob drove off first and the ball went up to a thousand feet in an arcing parabola and landed in a bunker one hundred and fifty yards down the fairway.

"That should bring rain," mused Henry quietly.

I would tell you at this juncture that our golf games are heavily larded with such pearls of wit. Jonesy hit a fair drive, Henry "topped" his and it scudded twenty yards and mine took

---

Tim Moxon was born in England in 1925 and was a bomber pilot in the Royal Air Force in World War II. He acted on the stages of London and had a small role as Commander Strangelove in *Doctor No,* a James Bond movie. Tim was a commercial pilot in Africa, then became a spray pilot in Africa, Jamaica and Panama. More recently, he has been flying his Hummingbird Air Tours in Montego Bay, Jamaica.

off from the toe of the club in a sizzling slice that narrowly missed a gentleman in plaid Bermuda shorts on the adjacent fairway.

I walked up the greensward and Jonesy put his hand on my shoulder. "Tell you a funny story. These two chaps were playing the ninth and a funeral goes past. One of them stands stiffly to attention and takes off his golf hat. "I say, old boy," says the other, "that's pretty decent to show respect for the dead." "Well," replies the first one, "least I can do really. You see, we were married for thirty-one years." A golfers joke, no doubt.

We parted company and Bernie and I went to retrieve my ball under the baleful scrutiny of the man in Bermuda shorts.

At the fifth tee of the Upton Country Club, it is necessary to drive across a fair-sized muddy lake festooned with water lilies. This particular hole is all the more unnerving on account of bystanders. There are never less than four or five small dark lads waiting for the inevitable and with a fair stock of golf balls which they have retrieved from the depths. We all put our drives into the lake and an air of lassitude crept into the game.

"Mebbe you let me have ah swing, boss?" Bernie propped my clubs against a coconut tree and took out a battered ball from the pocket.

"Sure, Bernie, go ahead. Hang on a minute fellows. Bernie's going to have a whack." He took the driver from the bag and balanced it in his hand like a Massai warrior checking his spear.

"Here, you'd better have one of these things," I called, proffering him a tee. "Makes the job a lot easier."

"Mi can't bother wid dem t'ings, boss." Bernie threw the ancient ball onto the grass and positioned it onto a dandelion with the clubhead. A quick glance toward the green, a sniff of the gentle breeze that rustled the palms and he swung.

The ball took off like a bullet, low and true, bounced once and settled three hundred yards down the fairway. For a long time none of us spoke as Bernie loped back to the bag and replaced the club. The silence was broken by Henry. 'Sheeeeit!" was all he said. Henry is an Australian and can be excused.

Bernie smiled, his crooked teeth very white in his black face.

"How many years have you been playing this game, Bernie?" I asked.

"Nevah had the bread fe a set of clubs . . . jist play sometime when I get the loan from one of de membahs."

On the far side of the lake, the four of us paid out fifty cents each to our followers and dropped the balls for a stroke. Passable approach shots brought us to within a hundred yards of the green. Bernie walked up to his ball in the middle of the fairway, shaded his eyes and looked at the green. He selected a number eight iron, replaced it and took out a number nine. After a laconic practice swing, he planted both feet, adorned by dirty white sneakers, square to the ball. One could speak of poetry in motion. Bernie just made me think of a big black cat. The ball soared high into the sky and landed within four feet of the pin, stopping dead. I didn't waste any time.

"Bernard, let me tell you something. The Jamaica Open is being played at Caymanas Park in three weeks. I'm putting your name down."

"Yu tink me have a chance, boss?" He lined up the putt and put it away for a birdie.

"Chance! Bernie, my friend, the way you are going, you're a candidate for the U.S. Open."

For the last four holes, I carried the clubs and Bernie played. The other three had stopped off at the bamboo bar on the 6th tee for a round of beers. Bernie played without fault. He flighted the ball round trees, blasted out of a bunker, and his approach game was incredible.

Over the next few weeks when I wasn't spraying bananas or running the hotel, I flew Bernie into Caymanas, which is near Kingston, and he practiced. He had new golf shoes. It made no difference. Bernie could have played in bare feet. I also bought him two pairs of golf slacks in light gray, terry cloth shirts and a tam-o'-shanter which he wore at a jaunty angle. We would drink a beer, eat a steak and he told me about his family. His father, whom he had never known, had taken off and Bernie had been brought up by his grandmother while his mother worked as a kitchen maid in one of the hotels.

The secretary of Caymanas, a jovial middle-aged Englishman, accepted the entry and the card. The afternoon before the tournament, Bernie and I climbed into the Cessna at the Ocho Rios airport. I had booked a couple of rooms at the historic Ferry Inn which is on the Kingston-Spanish Town road and within five minutes drive of the golf club.

It was six o'clock when we checked in and the bar was alive with the spectators of the morrow. I showed Bernie to his room and then we both went down the flagged stairway to the restaurant and ate a good meal.

"Early to bed tonight, sport," I mouthed over a piece of lobster. "Want you in spanking form and no mistake about that." My protege was looking smart and fit. Bernie was dining with his hat on. He loved that bloody hat.

At eight, after a last beer at the bar, I ushered him upstairs and said goodnight. "Nothing to be nervous about . . . just play your usual game and it's a cinch."

I slept fitfully as the discordant singing grew louder from the bar and awoke somewhat peevish to the smell of bacon frying and the clatter of dishes. I glanced at my watch — heavens! Nine o'clock and Bernie was due on the first tee at ten. I splashed water on my face, dressed hurriedly and went and knocked on Bernie's door. There was no reply. I knocked louder. "Bernie! Wake up — we're going to be late — just time for a quick breakfast."

I opened the door. Clothes were festooned over the floor and Bernie was lying across the bed in his underpants. One foot had a sock on it. He was snoring loudly. I shook him by the shoulders.

"Bernie! What the hell is wrong with you? Wake Up!" He rolled onto his back and opened one eye. It looked like a poached egg in a pool of blood. I took a glass, filled it with water from the washbasin and threw it in his face. I was taking a horrible chance.

"What di arse you do dat foh?" mumbled Bernie. He gazed up at me with the Cyclopian orb, rolled over and went back to sleep. "There is a tide in the affairs of men, which, taken at the flood, leads to fortune." This was not one of them. I hurried down to the bar and poured a large brandy, dashed back upstairs and held it under Bernie's nose. He sniffed once and, like a cat that is fond of fried fish, sat up and downed the drink in one gulp.

I got him into the shower, he dried and dressed with muted groans and seemed to tremble a lot.

"Is wha me hat den?" was all he said. "Some hunkies steal me hat."

"Screw the bloody hat," I said with some asperity, "let's go."

A large and friendly crowd was milling around the lounge of the country club when we arrived. Bernie was paired off with a young Lebanese banker and we shook hands and went out past the pool to the first tee. It may have been my imagination but Bernie seemed to be getting more sprightly — he even whistled a bar or two of "Island in the Sun."

The banker won the toss and drove off first; a solid shot and he stood back. Bernie threw the new golf ball down and took an almighty swing. There was a dull thud as he hit the ground a foot in front of the ball. There was a titter from the throng of spectators. I edged back and lost myself in the crowd, hurried back to the clubhouse and called a taxi from the airstrip.

I still see Bernie. There is no ill will and he waves and laughs when I drive past his rum shop on the Port Antonio road. Bernie was a "natural." The stuff of champions. Pity really.

The banker won the toss and drove off first—a solid shot and he stood back. Bernie threw the new golf ball down and took an almighty swing. There was a dull thud as he hit the ground a foot in front of the ball. There was a titter from the thronged spectators. I edged back and lost myself in the crowd. Hurried back to the clubhouse and called a taxi from the airstrip.

I still see Bernie. There is no ill will and he waves and laughs when I drive past his timbershop on the Port Antonio road. He was a "natural." The stuff of champions. Pity really.